A BEGINNER'S GUIDE TO STRUCTURAL EQUATION MODELING

Through its crystal clear explanations, this book is considered the most comprehensive introductory text to structural equation modeling (SEM). Noted for its thorough review of basic concepts *and* a wide variety of models, this book better prepares readers to apply SEM to a variety of research questions. Programming details and the use of algebra are kept to a minimum to help readers easily grasp the concepts so they can conduct their own analysis and critique related research. Featuring a greater emphasis on statistical power and model validation than other texts, each chapter features key concepts, examples from various disciplines, tables and figures, a summary, and exercises.

Highlights of the extensively revised fourth edition include:

- Uses different SEM software (not just Lisrel) including Amos, EQS, LISREL, Mplus, and R to demonstrate applications.
- Detailed introduction to the statistical methods related to SEM including correlation, regression, and factor analysis to maximize understanding (Chs. 1–6).
- The 5 step approach to modeling data (specification, identification, estimation, testing, and modification) is now covered in more detail and prior to the modeling chapters to provide a more coherent view of how to create models and interpret results (Ch. 7).
- More discussion of hypothesis testing, power, sampling, effect sizes, and model fit, critical topics for beginning modelers (Ch. 7).
- Each model chapter now focuses on one technique to enhance understanding by providing more description, assumptions, and interpretation of results, and an exercise related to analysis and output (Chs. 8–15).
- The use of SPSS AMOS diagrams to describe the theoretical models.
- The key features of each of the software packages (Ch. 1).
- Guidelines for reporting SEM research (Ch. 16).
- www.routledge.com/9781138811935 which provides access to data sets that can be used with any program, links to other SEM examples, related readings, and journal articles, and more.

Designed for introductory graduate courses in structural equation modeling, factor analysis, advanced, multivariate, or applied statistics, quantitative techniques, or statistics II taught in psychology, education, business, and the social and healthcare sciences, this practical book also appeals to researchers in these disciplines. Prerequisites include an introduction to intermediate statistics that covers correlation and regression principles.

RANDALL E. SCHUMACKER is a Professor of Educational Research at The University of Alabama, where he teaches courses in structural equation modeling.

RICHARD G. LOMAX is a Professor in the Department of Educational Studies at The Ohio State University.

"This book gives me what I need to get the students going, it well-grounds them in the basics, and it sets up a number of advanced topics that I can elaborate on. ... This book is appropriate for an introductory graduate course on structural equation modeling, or for professionals who want to learn SEM." **–Craig Parks, Washington State University, USA**

A BEGINNER'S GUIDE TO STRUCTURAL EQUATION MODELING

Fourth Edition

Randall E. Schumacker and Richard G. Lomax

Routledge
Taylor & Francis Group

NEW YORK AND LONDON

This edition published 2016
by Routledge
711 Third Avenue, New York, NY 10017

and by Routledge
2 Park Square, Milton Park, Abingdon, Oxon, OX14 4RN

Routledge is an imprint of the Taylor & Francis Group, an informa business

First edition published by Lawrence Erlbaum Associates, 1996
Third edition published by Taylor & Francis, 2010

Library of Congress Cataloguing in Publication Data
Schumacker, Randall E.
 A beginner's guide to structural equation modeling / Randall E. Schumacker,
 Richard G. Lomax. – Fourth edition.
  pages cm
 Includes bibliographical references and index.
 1. Structural equation modeling. 2. Regression analysis. 3. Social
 sciences–Statistical methods. I. Lomax, Richard G. II. Title.
 QA278.S36 2016
 519.5′3–dc23   2015021490

ISBN: 978-1-138-81190-4 (hbk)
ISBN: 978-1-138-81193-5 (pbk)
ISBN: 978-1-315-74910-5 (ebk)

Typeset in Times New Roman
by Out of House Publishing

Printed and bound in the United States of America by Sheridan Books, Inc. (a Sheridan Group Company).

Dedicated to Our Colleagues
and
SEM Researchers Worldwide

PREFACE

APPROACH

This book presents a basic introduction to structural equation modeling (SEM). Readers will find that we have followed our tradition of keeping examples rudimentary and easy to follow. The reader is provided with a review of correlation and covariance, followed by multiple regression, path, and factor analysis in order to better understand the building blocks of SEM. We then describe the basic steps a researcher takes in conducting structural equation modeling: model specification, model identification, model estimation, model testing, and sometimes model modification.

The different types of structural equation models are presented in individual chapters. Our approach is to focus the researcher on the scope and application of each type of model. The book was written to provide both a conceptual and application-oriented approach. Each chapter covers basic concepts, principles, and practice, then utilizes SEM software to provide a meaningful example. Chapters 8 to 15 provide the different SEM model examples with an exercise to permit further practice. The SEM model examples in the book do not require complicated programming skills, nor does the reader need an advanced understanding of statistics and matrix algebra to understand the model applications. We have provided an appendix on matrix operations for the interested reader. We encourage the understanding of the matrices used in SEM models, especially for some of the more advanced SEM models you will encounter in the research literature. It is wise when publishing your research to express your SEM model in matrix form.

CONTENT COVERAGE

Our main goal in this fourth edition is to provide a basic introduction to SEM model analyses, offer SEM model examples, and provide resources to better understand SEM research. These goals are supported by the conceptual and applied examples contained in the book chapters, and a website with data, programs, and Internet links to related documents and journal articles. We have also provided journal article references at the end of most chapters. The last chapter includes a discussion of SEM guidelines for reporting SEM research.

We have organized the chapters based on reviewer feedback. We begin with an introduction to SEM (what it is, some history, why conduct it, and what software is available), followed by chapters on data entry and editing issues, and correlation. These early chapters are critical to understanding how missing data, non-normality, scale of measurement, non-linearity, outliers, and restriction of range in scores affect SEM analysis.

We proceed with multiple regression models, path models, and factor models. We explore the basics of model fit in multiple regression using the R^2 value, as well as the limitations of multiple regression models (saturated model, measurement error, direct effects only). We continue with path models, where we discuss the drawing conventions for SEM models. In path models, we also learn about direct and indirect effects, but continue with knowing that observed variables in path models contain measurement error. An important lesson with path models is the decomposition of the variance–covariance (correlation) matrix. This provides the essential understanding of how a theoretical model implies all or most of the sample variance–covariance relations amongst the observed variables. It also introduces the chi-square statistic as a model fit between the sample matrix and the model-implied matrix. Finally, we introduce exploratory and confirmatory factor analysis. Our goal is to demonstrate how multiple measures (observed variables) share variance in defining a latent variable (construct). It introduces the concept that measurement error is removed from observed variables. This also permits a discussion of latent variables that are used in structural equation modeling.

We have set apart Chapter 7 to discuss in detail SEM principles and practice. Specifically, we detail the SEM modeling steps. The types of model-fit criteria are presented, the testing of parameter estimates in models, and the concepts of saturated and independent models are also explained. This chapter is also important because it discusses hypothesis testing and the related issues of power, sample size, and effect size in SEM. We also express the need to report additional fit measures with the chi-square test of model fit. The additional fit measures are categorized by model fit, model parsimony, or model comparison strategies when testing SEM

ABOUT THE AUTHORS

RANDALL E. SCHUMACKER is a Professor of Educational Research at The University of Alabama, where he teaches courses in structural equation modeling, multivariate statistics, multiple regression, and program evaluation. His research interests are varied, including modeling interaction in SEM, robust statistics (normal scores, centering and variance inflation factor issues), specification search issues as well as measurement model issues related to estimation, mixed-item formats, and reliability. He has taught several international and national workshops on structural equation modeling.

Randall has written and co-edited several books, including *A Beginner's Guide to Structural Equation Modeling* (third edition); *Advanced Structural Equation Modeling: Issues and Techniques*; *Interaction and Non-Linear Effects in Structural Equation Modeling*; *New Developments and Techniques in Structural Equation Modeling*; *Understanding Statistical Concepts Using S-PLUS*; *Understanding Statistics Using R*; *Learning Statistics Using R*; and *Using R with Multivariate Statistics*.

Randall has published in several journals including *Academic Medicine*, *Educational and Psychological Measurement*, *Journal of Applied Measurement*, *Journal of Educational and Behavioral Statistics*, *Journal of Research Methodology*, *Multiple Linear Regression Viewpoints*, and *Structural Equation Modeling*. He has served on the editorial boards of numerous journals and is a member of the American Educational Research Association, Past-President of the Southwest Educational Research Association, and Emeritus Editor of *Structural Equation Modeling* and *Multiple Linear Regression Viewpoints*.

Dr. Schumacker was the 1996 recipient of the *Outstanding Scholar Award*, and the 1998 recipient of the *Charn Oswachoke International Award*. In 2010, he

ABOUT THE AUTHORS

RANDALL E. SCHUMACKER is a Professor of Educational Research at The University of Alabama, where he teaches courses in structural equation modeling, multivariate statistics, multiple regression, and program evaluation. His research interests are varied, including modeling interaction in SEM, robust statistics (normal scores, centering and variance inflation factor issues), specification search issues as well as measurement model issues related to estimation, mixed-item formats, and reliability. He has taught several international and national workshops on structural equation modeling.

Randall has written and co-edited several books, including *A Beginner's Guide to Structural Equation Modeling* (third edition); *Advanced Structural Equation Modeling: Issues and Techniques*; *Interaction and Non-Linear Effects in Structural Equation Modeling*; *New Developments and Techniques in Structural Equation Modeling*; *Understanding Statistical Concepts Using S-PLUS*; *Understanding Statistics Using R*; *Learning Statistics Using R*; and *Using R with Multivariate Statistics*.

Randall has published in several journals including *Academic Medicine, Educational and Psychological Measurement, Journal of Applied Measurement, Journal of Educational and Behavioral Statistics, Journal of Research Methodology, Multiple Linear Regression Viewpoints*, and *Structural Equation Modeling*. He has served on the editorial boards of numerous journals and is a member of the American Educational Research Association, Past-President of the Southwest Educational Research Association, and Emeritus Editor of *Structural Equation Modeling* and *Multiple Linear Regression Viewpoints*.

Dr. Schumacker was the 1996 recipient of the *Outstanding Scholar Award*, and the 1998 recipient of the *Charn Oswachoke International Award*. In 2010, he

Chapter 6 further explains the content related to constructing latent variables, and the relationship between exploratory and confirmatory factor analysis. The book now illustrates the theoretical SEM models in Chapters 8 to 15 by using SPSS AMOS diagrams. The diagrams help the beginner to visually conceptualize the theoretical models.

The past several years have seen an explosion in SEM coursework, books, websites, and training courses. We are proud to have been considered a starting point for many beginners to SEM. We hope you find that this fourth edition provides the updated information a beginner needs regarding selection of software, trends, and topics in SEM today.

ACKNOWLEDGMENTS

The fourth edition of this book represents more than 35 years of interacting with our colleagues and students who use structural equation modeling. As before, we are most grateful to the pioneers in the field of structural equation modeling, particularly to Karl Jöreskog, Dag Sörbom, Peter Bentler, James Arbuckle, and Linda and Bengt Muthèn. These individuals have developed and shaped the advances in the SEM field as well as the content of this book, plus provided SEM researchers with software programs, data sets, and examples. We are grateful to their companies (IBM-Amos, EQS, LISREL, and Mplus) for providing software to use in our chapter examples. We would also like to thank the reviewers who provided valuable feedback on the initial revision plan, including Brian A. Lawton, George Mason University; Jos Schijns, Open Universiteit, The Netherlands; Linda Shanock, University of North Carolina at Charlotte; Hongwei Yang, University of Kentucky; and one anonymous reviewer.

This book was made possible through the encouragement of Debra Riegert at Routledge/Taylor & Francis who insisted it was time for a fourth edition. We wish to thank her for coordinating all of the activity required to get a book into print.

Randall E. Schumacker, University of Alabama

Richard G. Lomax, Ohio State University

since writing the first edition of the book because researchers either have a preference and experience with a certain SEM package, or do not have access to the software. Today, popular commercial SEM software packages have free demonstration versions, and there are free software packages available, which are described in Chapter 1. We have therefore provided author-created SEM software programs and data sets in ASCII format, which permits the researcher to open them without having the software installed. This can be done using *Notepad*. It permits a quick look at the code and syntax used by the SEM software. A researcher can also copy and paste the contents into the SEM software to access the data set and run the program if so desired. A student or researcher will quickly learn that the program syntax for most SEM software programs only requires a few simple commands.

An important new feature of the fourth edition is the creation of an enhanced website. The website at www.routledge.com/9781138811935/ provides easy access to SEM resources. We have included the data sets used in the book, in ASCII format so they can be used in any SEM program. Given the proliferation of SEM over the past several decades, we have also included website links to other examples and information about the particular SEM examples we discuss in the book. In some cases, we have also provided publicly available web links to Adobe PDF files that further discuss or present information about the SEM examples we present in the chapters. This gives readers additional insight into the SEM approach.

The fourth edition updates and continues to provide key references. Where important, new book and journal article references are provided. We have provided this information as follow-up to our basic discussion. SEM methods, applications, techniques, issues, and trends have advanced over the past several decades. This edition is intended to provide introductory information to get a beginner started understanding and using SEM. The only prerequisites are that you have a basic background in intermediate statistics that covers correlation and regression principles. The references should help direct the reader to additional resources. We have expanded our discussion in Chapter 2 of the data editing and screening issues related to data input format types, start values, missing data, normality, invariance, and positive definite matrices. We have expanded discussion of correlation and covariance issues (Chapter 3), regression types (Chapter 4), direct/indirect effects and matrix decomposition (Chapter 5), types of factor analysis techniques, and issues related to number of factors, sample size, rotation, and factor scores, exploratory versus confirmatory factor analysis, measurement invariance, and correlated error terms in factor models (Chapter 6). The enhanced discussion in Chapters 3 to 5 further clarifies the role correlation plays in SEM modeling, the basis for understanding how a path model decomposes a correlation matrix, and an understanding of direct and indirect effects in a model.

interpretation. We strongly encourage a thorough understanding of the SEM basics in Chapter 7. We have also expanded the coverage of the different SEM models to provide more description of the models, assumptions, and how to interpret results in Chapters 8 to 15. Each of these chapters now focuses on a single SEM model application. This permits a more focused in-depth discussion of the rationale and approach taken to the SEM analysis. The chapters are reorganized to better flow with an increased knowledge and capability in SEM modeling. Each chapter concludes with an exercise that the student or researcher can run, which permits an enhanced understanding of the software, analysis, output, and interpretation. Finally, Chapter 16 was revised to provide guidelines and suggestions for reporting SEM research.

We have included a variety of SEM examples using different SEM software to further an understanding of SEM research applications. The SEM software programs include Amos, EQS, LISREL, Mplus, and R examples. We have used commercially available software in some examples, and also used the free demonstration versions in some chapter examples. The demonstration versions can be downloaded from the websites for Amos, EQS, LISREL, Mplus, and R. Please be aware that their demonstration versions of software, although free, are limited and do not contain all of the features provided by a purchased version. Given the advances in SEM software over the past decade, you should expect updates and patches to software packages and therefore become familiar with any new features, as well as explore the excellent library of examples and help materials available in your SEM software.

The fourth edition offers several changes and additions not in the previous editions. The first chapter provides website links for many of the commercial and free SEM software packages. A brief description of the key features in each software package is also provided. The next five chapters cover in more detail the basic understanding of correlation, regression, path, and factor analysis necessary in SEM practice. We have included a detailed discussion of the importance of the determinant of a matrix, and the resulting eigenvalues and eigenvectors. A researcher will learn that a zero or negative determinant of a matrix results in a non-positive definite matrix, which causes SEM programs to stop running and give an error message.

Our discussion of factor analysis now includes exploratory and confirmatory methods. This addresses the key issue of the difference between pattern and structure matrices in factor analysis. We further discuss how latent variable scores are computed from confirmatory factor models.

Another new feature in the fourth edition is the use of different SEM software programs when testing the different SEM models. This has always been a challenge models. This is a critical chapter that should be studied in depth before proceeding with the remaining chapters in the book.

Chapters 8 to 15 provide different types of SEM models. The SEM steps of model specification, identification, estimation, testing, and modification are explained for each type of SEM model. These chapters provide a basic introduction to the rationale and purpose of each SEM model. The chapter exercises permit further practice and understanding of how each SEM model is analyzed.

Theoretical SEM models are present in many academic disciplines, and therefore we have tried to present a variety of different models that can be formulated and tested. This fourth edition attempts to cover SEM models and applications a student or researcher may use in psychology, education, sociology, business, medicine, political science, and the biological sciences. Our focus is to provide the basic concepts, principles, and practice necessary to test the theoretical models. We hope you become more familiar with structural equation modeling after reading the book, and use SEM in your own research.

NEW TO THE FOURTH EDITION

The first edition of this book was one of the first books published on SEM, while the second edition greatly expanded knowledge of advanced SEM models. Since that time, we have had considerable experience utilizing the book in class with our students. The third edition was greatly expanded and represented a more useable book for teaching SEM. The fourth edition returns to a basic introductory book with concepts, principles, and practice related to how different SEM models are analyzed.

This was accomplished by reorganizing the book to more closely reflect how students learn about and use SEM techniques. We provide more background information in Chapters 1 and 2 and a discussion of methods related to SEM (correlation, regression, path analysis, and factor analysis) in Chapters 3 to 6. In addition, Chapter 7 now covers how to develop and test SEM models prior to the individual SEM modeling chapters so as to re-emphasize the steps a student takes in SEM modeling. We also provide more information on hypothesis testing, power, sample size, and effect size in Chapter 7 and show how the RMSEA (root mean square error of approximation) model fit value can be used to state the null and alternative hypothesis. The important issues of power, sample size, and effect size are also discussed with examples and key references. This has been a critical area of understanding for many beginning SEM modelers. In addition, we have simplified the discussion of fit measures, which ones to report, and their

CONTENT COVERAGE

Our main goal in this fourth edition is to provide a basic introduction to SEM model analyses, offer SEM model examples, and provide resources to better understand SEM research. These goals are supported by the conceptual and applied examples contained in the book chapters, and a website with data, programs, and Internet links to related documents and journal articles. We have also provided journal article references at the end of most chapters. The last chapter includes a discussion of SEM guidelines for reporting SEM research.

We have organized the chapters based on reviewer feedback. We begin with an introduction to SEM (what it is, some history, why conduct it, and what software is available), followed by chapters on data entry and editing issues, and correlation. These early chapters are critical to understanding how missing data, non-normality, scale of measurement, non-linearity, outliers, and restriction of range in scores affect SEM analysis.

We proceed with multiple regression models, path models, and factor models. We explore the basics of model fit in multiple regression using the R^2 value, as well as the limitations of multiple regression models (saturated model, measurement error, direct effects only). We continue with path models, where we discuss the drawing conventions for SEM models. In path models, we also learn about direct and indirect effects, but continue with knowing that observed variables in path models contain measurement error. An important lesson with path models is the decomposition of the variance–covariance (correlation) matrix. This provides the essential understanding of how a theoretical model implies all or most of the sample variance–covariance relations amongst the observed variables. It also introduces the chi-square statistic as a model fit between the sample matrix and the model-implied matrix. Finally, we introduce exploratory and confirmatory factor analysis. Our goal is to demonstrate how multiple measures (observed variables) share variance in defining a latent variable (construct). It introduces the concept that measurement error is removed from observed variables. This also permits a discussion of latent variables that are used in structural equation modeling.

We have set apart Chapter 7 to discuss in detail SEM principles and practice. Specifically, we detail the SEM modeling steps. The types of model-fit criteria are presented, the testing of parameter estimates in models, and the concepts of saturated and independent models are also explained. This chapter is also important because it discusses hypothesis testing and the related issues of power, sample size, and effect size in SEM. We also express the need to report additional fit measures with the chi-square test of model fit. The additional fit measures are categorized by model fit, model parsimony, or model comparison strategies when testing SEM launched the *DecisionKit* App for the iPhone and iPad, which can assist researchers in making decisions about which measurement, research design, or statistic to use in their research projects. In 2011, he received the *Apple iPad Award*, and in 2012, he received the *CIT Faculty Technology Award*. In 2013, he received the *McCrory Faculty Excellence in Research Award* from the College of Education at the University of Alabama. In 2014, Dr. Schumacker was the recipient of the *Structural Equation Modeling Service Award* at the American Educational Research Association, where he founded the Structural Equation Modeling Special Interest Group. He can be contacted at The University of Alabama, College of Education, P.O. Box 870231, 316 Carmichael Hall, Tuscaloosa, AL 35487-0231, USA or by e-mail at rschumacker@ua.edu.

RICHARD G. LOMAX is a Professor in the Department of Educational Studies at The Ohio State University. He received his Ph.D. in Educational Research Methodology from the University of Pittsburgh. His research focuses on models of literacy acquisition, multivariate statistics, analysis of variance, and assessment. He has thrice served as a Fulbright Scholar and is a Fellow of the American Educational Research Association. Richard can be contacted at The Ohio State University, College of Education and Human Ecology, 153 Arps Hall, 1945 N. High Street, Columbus, OH 43210, USA or by e-mail at lomax.24@osu.edu.

## Chapter 1

INTRODUCTION

CHAPTER CONCEPTS

What is structural equation modeling?
History of structural equation modeling
Why conduct structural equation modeling?
Structural equation modeling software

WHAT IS STRUCTURAL EQUATION MODELING?

Structural equation modeling (SEM) depicts relations among observed and latent variables in various types of theoretical models, which provide a quantitative test of a hypothesis by the researcher. Basically, various theoretical models are hypothesized and tested in SEM. The SEM models hypothesize how sets of variables define constructs and how these constructs are related to each other. In SEM, the construct is called a latent variable. For example, an educational researcher might hypothesize that a student's home environment influences her later achievement in school. A marketing researcher may hypothesize that consumer trust in a corporation leads to increased product sales for that corporation. A health care professional might believe that a good diet and regular exercise reduces the risk of a heart attack.

In each example, based on theory and empirical research, the researcher wants to test whether a set of variables define the constructs that are hypothesized to be related in a certain way. The goal of SEM is to test whether the theoretical model is supported by sample data. If the sample data support the theoretical model, then the hypothesized relations exist amongst the constructs. If the sample data do not support the theoretical model, then either an alternative model will need to be specified and tested, or another theoretical model hypothesized and tested. Consequently, SEM tests theoretical models using the scientific method of hypothesis testing to advance our understanding of the complex relations among constructs.

SEM can test various types of theoretical models. The first types discussed in this book include regression, path, and confirmatory factor models (CFA), which form the basis for understanding the many different types of SEM models. The regression models use observed variables, while path models can use either observed variables or latent variables. CFA models by definition use observed variables to define latent variables; however, second-order CFA models test relations using additional latent variables. Therefore, these two types of variables, *observed variables* and *latent variables,* are used depending upon the type of SEM model.

Latent variables (constructs or factors) are variables that are not directly observed or measured. Latent variables are indirectly observed or measured, and hence are inferred from a set of observed variables that we actually measure using tests, surveys, scales, and so on. For example, intelligence is a latent variable that represents a psychological construct. The confidence of consumers in American business is another latent variable, one representing an economic construct. The physical condition of adults is a third latent variable, one representing a health-related construct.

The observed variables (measured or indicator) are a set of variables that we use to define or infer the latent variable or construct. For example, the Wechsler Intelligence Scale for Children—Revised (WISC-R) is an instrument that produces a measured variable (scores), which is used to infer the construct of a child's intelligence. Additional indicator variables of intelligence tests would be used to indicate or define the construct of intelligence (latent variable). The Dow Jones index is a standard measure of the American corporate economy construct. Other indicator variables could include gross national product, retail sales, and export sales. Blood pressure is an indicator of a health-related latent variable that could be defined as *Fitness*. Other indicator variables could be exercise and diet. Researchers use several indicator variables to define a latent variable. The CFA measurement model tests whether the indicator variables are a good fit in defining the latent variable. The SEM structural model then tests the hypothesized relations amongst the latent variables.

Latent variables are defined as either *independent variables* or *dependent variables*. An independent variable is a variable that is not manipulated or influenced by any other variable in the model. A dependent variable is a variable that is influenced by other variables in the model. The researcher specifies the independent and dependent variables. The educational researcher hypothesizes that a student's home environment (independent latent variable) influences school achievement (dependent latent variable). The marketing researcher believes that consumer trust in a corporation (independent latent variable) leads to increased product sales (dependent latent variable). The health care professional wants to determine whether a good diet, regular exercise, and physiology (independent latent variable) influences the frequency of heart attacks (dependent latent variable).

The basic SEM models (regression, path, and CFA) illustrate the use of observed variables and latent variables which are defined as independent or dependent in the model. A regression model consists solely of observed variables where a single dependent observed variable is predicted or explained by one or more independent observed variables; for example, a parent's income level (independent observed variable) is used to predict his or her child's achievement score (dependent observed variable). A path model can also be specified entirely with observed variables, but the flexibility allows for multiple independent observed variables and multiple dependent observed variables; for example, export sales, gross national product, and NASDAQ index (independent observed variables) influence consumer trust and consumer spending (dependent observed variables). Path models test more complex models than regression models, include direct and indirect effects, and can include latent variables. For example, home environment, school environment, and relations with peers (independent latent variables) can explain student achievement and student–teacher relations (dependent latent variables). Confirmatory factor models consist of observed variables that are hypothesized to measure both the independent and dependent latent variables; for example, diet, exercise, and physiology are observed measures of the independent latent variable, *Fitness*, while blood pressure, cholesterol, and stress are observed measures of the dependent latent variable, *Heart Attack Proneness*. As another example, an independent latent variable (home environment) influences a dependent latent variable (achievement), where both types of latent variables are measured by multiple observed (indicator) variables.

HISTORY OF STRUCTURAL EQUATION MODELING

To discuss the history of structural equation modeling, we explain the following four types of related models and their chronological order of development: regression, path, confirmatory factor, and structural equation modeling.

The first model involves linear regression models that use a correlation coefficient and the least squares criterion to compute regression weights. Regression models were made possible because Karl Pearson created a formula for the correlation coefficient in 1896 that provided an index for the relation between two variables (Pearson, 1938). The regression model permits the prediction of dependent observed variable scores (*Y scores*), given a linear weighting of a set of independent observed scores (*X scores*). The linear weighting of the independent variables is done using regression coefficients, which are determined based on minimizing the sum of squared residual error values. The selection of the regression weights is therefore based on the *Least Squares Criterion*. The mathematical basis for the linear regression model is found in basic algebra. Regression analysis provides a test of a theoretical model that may be useful for prediction, for example, admission to graduate school or budget projections. Delucchi (2006), for example, used regression analysis to predict student exam scores in statistics (dependent variable) from a series of collaborative learning group assignments (independent variables). The results provided some support for collaborative learning groups improving statistics of exam performance, although not for all tasks.

Some years later, Charles Spearman (1904, 1927) used the correlation coefficient to determine which items correlated or went together to create a measure of general intelligence. His basic idea was that if a set of items correlated or went together, individual responses to the set of items could be summed to yield a score that would measure or define a construct. Spearman was the first to use the term *factor analysis* in defining a two-factor construct for a theory of intelligence. D. N. Lawley and L. L. Thurstone in 1940 further developed applications of factor models, and proposed instruments (sets of items) that yielded observed scores from which constructs could be inferred. Most of the aptitude, achievement, and diagnostic tests, surveys, and inventories in use today were created using factor analytic techniques. The term *confirmatory factor analysis* (CFA) is used today based in part on earlier work by Howe (1955), Anderson and Rubin (1956), and Lawley (1958). The CFA method was more fully developed by Karl Jöreskog in the 1960s to test whether a set of items defined a construct. Jöreskog completed his dissertation in 1963, published the first article on CFA in 1969, and subsequently helped develop the first CFA software program. Factor analysis has been used for more than 100 years to create measurement instruments in many academic disciplines. CFA today uses observed variables derived from measurement instruments to test the existence of a theoretical construct. Goldberg (1990) used CFA to confirm the *Big Five* model of personality. His five-factor model of extraversion, agreeableness, conscientiousness, neuroticism, and intellect was confirmed through the use of multiple indicator variables for each of the five hypothesized constructs.

The path model was developed by Sewell Wright (1918, 1921, 1934), a biologist. Path models use correlation coefficients and multiple regression equations to model more complex relations amongst observed variables. The first application of path models dealt with animal behavior. Unfortunately, path analysis was largely overlooked until econometricians reconsidered it in the 1950s as a form of simultaneous equation modeling (Wold, 1954) and sociologists rediscovered it in the 1960s (Duncan, 1966) and 1970s (Blalock, 1972). A path analysis involves solving a set of simultaneous regression equations that theoretically establish the relations amongst the observed variables in the path model. Parkerson et al. (1984) conducted a path analysis to test Walberg's theoretical model of educational productivity for fifth- through eighth-grade students. The relations amongst the following variables were analyzed in a single model: home environment, peer group, media, ability, social environment, time on task, motivation, and instructional strategies. All of the hypothesized paths among those variables were shown to be statistically significant, providing support for the educational productivity path model.

Structural equation models (SEM) combine path models and confirmatory factor models when establishing hypothesized relations amongst latent variables. The early development of SEM models was due to Karl Jöreskog (1969, 1973), Ward Keesling (1972), and David Wiley (1973); the approach was initially known as the JKW model, but became known as the *li*near *s*tructural *rel*ations model (LISREL) with the development of the first software program, *LISREL*, in 1973.

Jöreskog and van Thillo originally developed the LISREL software program at the Educational Testing Service (ETS) using a matrix command language that used Greek and matrix notation. The first publicly available version, LISREL III, was released in 1976. By 1993, LISREL8 was released; it introduced the SIMPLIS (SIMPle LISrel) command language in which equations were written using variable names. In 1999, the first interactive version of LISREL was released. LISREL8 introduced the dialog box interface using pull-down menus and point-and-click features to develop models. The path diagram mode permitted drawing a program to develop models. LISREL9 has since been released with new features to address categorical and continuous variables. Karl Jöreskog was recognized by Cudeck, DuToit, and Sörbom (2001) who edited a Festschrift in honor of his contributions to the field of structural equation modeling. Their volume contains chapters by scholars who addressed the many topics, concerns, and applications in the field of structural equation modeling today, including milestones in factor analysis; measurement models; robustness, reliability, and fit assessment; repeated measurement designs; ordinal data; and interaction models.

The field of structural equation modeling across all disciplines has expanded since 1994. Hershberger (2003) found that between 1994 and 2001 the number of journal articles concerned with SEM increased, the number of journals publishing SEM research increased, SEM became a popular choice amongst multivariate methods, and the journal *Structural Equation Modeling* became the primary source for technical developments in structural equation modeling, and continues so today. SEM research articles are now more prevalent than ever in professional journals of several different academic disciplines (medicine, psychology, business, education, etc.).

WHY CONDUCT STRUCTURAL EQUATION MODELING?

Why is structural equation modeling popular? There are at least four major reasons for the popularity of SEM. First, researchers are becoming more aware of the need to use multiple observed variables to investigate their area of scientific inquiry. Basic statistical methods only utilize a limited number of independent and dependent variables, and thus do not test theoretical relations amongst multiple variables. The use of a small number of variables to understand complex phenomena is limited. For instance, the bivariate correlation is not sufficient for examining prediction when using multiple variables in a regression equation. In contrast, structural equation modeling permits relations amongst multiple variables to be modeled and statistically tested. SEM techniques are therefore a more preferred method to confirm (or disconfirm) theoretical models.

Second, a greater recognition has been given to the validity and reliability of observed scores from measurement instruments. Specifically, measurement error has become a major issue in many disciplines, although historically, statistical analysis of data has ignored measurement error in the analysis of data. Structural equation modeling techniques explicitly take measurement error into account when statistically analyzing data. SEM analysis includes latent and observed variables with their associated measurement error terms in the many different SEM models.

Third, structural equation modeling has matured over the past 40 years, especially the software programs and the ability to analyze more advanced theoretical SEM models. For example, group differences in theoretical models can be tested with multiple-group SEM models. The analysis of educational data collected at more than one level (school districts, schools, and teachers) with student data is now possible using multi-level SEM models. SEM models are no longer limited to linear relations; interaction terms can now be included in an SEM model so that main effects and interaction effects can be tested. The improvement in SEM software has led to advanced SEM models and techniques, which have provided researchers with an increased capability to analyze sophisticated theoretical models of complex phenomena.

Fourth, SEM software programs have become increasingly user-friendly. Until 1993, SEM modelers had to input the program syntax for their models using Greek and matrix notation. At that time, many researchers sought help because of the complex programming and knowledge of the SEM syntax that was required. Today, SEM software programs are easier to use and contain features similar to other Windows-based software packages, for example, pull-down menus, data spreadsheets, and a simple set of commands. However, the user-friendly SEM software, which permitted more access by researchers, comes with concerns about proper usage. Researchers need the pre-requisite training in statistics and specifically in SEM modeling. Fortunately, there are courses, workshops, and textbooks one can acquire to avoid mistakes and errors in analyzing sophisticated theoretical models using SEM. Which SEM software program you choose may also influence your level of expertise and what type of SEM models can be analyzed.

STRUCTURAL EQUATION MODELING SOFTWARE

Structural equation modeling can be easily understood if the researcher has a grounding in basic statistics, correlation, regression, and path analysis. Some SEM software programs provide a pull-down menu with these capabilities, while others come included in a statistics package where they can be computed. Although the LISREL program was the first SEM software program, other software programs have subsequently been developed since the mid-1980s. These include EQS, Mplus, Mx, R, Proc Calis (SAS), AMOS (SPSS), Sepath (Statistica), and SEM (STATA), to name a few. These software programs are each unique in their own way, with some offering specialized features for conducting different SEM applications. Many of these SEM software programs provide statistical analysis of raw data (means, correlations, missing data conventions), provide routines for handling missing data and detecting outliers, generate the program's syntax, diagram the model, and provide for import and export of data and figures of a theoretical model. Also, many of the SEM software programs come with sets of data and program examples that are clearly explained in their user guides. Many of these software programs have been reviewed in the journal *Structural Equation Modeling*.

The pricing information for SEM software varies depending on individual, group, or site license arrangements; corporate versus educational settings; and even whether one is a student or faculty member. Furthermore, newer versions and updates necessitate changes in pricing. We are often asked to recommend a software package to a beginning SEM researcher; however, given the different individual needs of researchers and the multitude of different features available in these programs, we are not able to make such a recommendation. Ultimately the decision depends upon the researcher's needs and preferences.

The many different SEM software programs are either stand-alone or part of a statistical package. Most SEM programs run in the Windows environment, but not all. Some have demo versions, while others are free. Several SEM software programs are described below in alphabetical order with links to their websites. These links are also on the book website.

AMOS (SPSS)

AMOS is an add-on purchase in the IBM SPSS statistical package. It uses diagramming tools to draw the SEM model, then links the model and variables to an SPSS data set. Some of its features are the ability to drag and drop the variable names into the model from the SPSS data set, the ability to export the SEM model diagram to Word via a clipboard tool, the analysis of multiple-group models, and the ability to conduct a specification search for a better model. More information about AMOS can be found at: www-03.ibm.com/software/products/en/spss-amos/

EQS: Stand-alone Software

EQS software comes with a program manual, user's guide, and free technical support. Some of its features are new and improved missing data procedures, reliability coefficients for factor models, Satorra–Bentler robust corrected standard errors in the presence of non-normal and missing data issues, multi-level modeling options, bootstrap simulation capabilities, latent growth curve modeling, and a simple set of syntax commands. More information about EQS can be found at: www.mvsoft.com/products.htm

JMP: Stand-alone Software / SAS Interface

JMP is a graphical user interface software product for Windows PC-based applications. It is a statistical, graphical, and spreadsheet-based software product that does not necessarily require writing program syntax. SEM runs in JMP version 10.0.2 for Macintosh, Windows XP, and Windows 7 operating systems. SAS Structural Equation Modeling for JMP is a new application that enables researchers to use SAS and JMP to draw models by using an interface that is built on the SAS/STAT® CALIS procedure. JMP Pro conducts partial least squares (PLS) modeling. However, you can write and submit SAS code. SAS created JMP in 1989 to empower scientists and engineers to explore data visually. Since then, JMP has grown from a single product into a family of statistical discovery tools, each one tailored to meet specific needs. All of the software is visual, interactive, comprehensive, and extensible. More information about JMP is available at: www.jmp.com/software/

LISREL: Stand-alone Software

LISREL has evolved from a matrix-only software (LISREL9) to a simple set of syntax commands (SIMPLIS). You can obtain output in matrix format or obtain the basic results. The software comes with a free demo version, user guides, sets of data, and examples for the many different types of SEM models with a help menu to search for information and explanations of analysis and procedures. PRELIS is a pre-processor for the data to check variable missing data, non-normality (skewness and kurtosis), and data type (nominal, ordinal, interval/ratio). The pull-down menu in SIMPLIS also provides for data imputation (presence of missing data); classical and ordinal factor analysis; censored, logistic, and probit regression; bootstrapping, and output options to save files such as the asymptotic variance–covariance matrix. Other features on the pull-down menu include multi-level modeling and survey data analysis using the general linear model.

LISREL9 now displays the determinant and the maximum and minimum eigenvalues. The determinant is computed as the product of the eigenvalues. The determinant is printed as the generalized variance in a matrix. In matrix algebra, the eigenvalues are canonical r squared values. In principal components analysis (PCA), the sum of eigenvalues is the amount of variance that can be explained. In PCA, the sum of 1's in the correlation matrix equals the total variance to decompose (also the number of variables in the matrix). This is different from factor analysis.

The total variance is the sum of the diagonal elements of S (trace of matrix), and the generalized variance is the determinant of S which equals the product of all the eigenvalues of S. The largest and smallest eigenvalues of S are printed in LISREL9. The condition number is the square root of the ratio of the largest and smallest eigenvalue. A small condition number indicates multicollinearity in the data. If the condition number is very small LISREL gives a warning. This might indicate that one or more variables are linear or nearly linear combinations of other variables.

Structural equation modeling (SEM) was introduced as a way to test a theoretical model analyzing the covariance or correlation matrix. Initially, a researcher would read the matrix into LISREL and estimate the model parameters using the maximum likelihood estimation method. If raw data were available without missing values, one could also use PRELIS first to estimate an asymptotic covariance matrix to obtain robust estimates of standard errors and chi-squares. For examples based on covariance and correlation matrices, see the LISREL user guide (Jöreskog & Sörbom, 2006).

Structural equation modeling can also be conducted using raw data. If raw data are available in a LISREL system file, LISREL9 will automatically perform robust estimation of standard errors and chi-square goodness-of-fit measures under non-normality. It is no longer necessary to estimate an asymptotic covariance matrix with PRELIS and read this into your program. The estimation of the asymptotic covariance matrix and the model is now done in LISREL9. If the data contain missing values, LISREL9 will automatically use FIML to estimate the model. Alternatively, users may choose to impute the missing values by EM or Markov chain Monte Carlo (MCMC) and estimate the model based on the imputed data. All this works for both continuous and ordinal variables. More information about LISREL can be found at: www.ssicentral.com/lisrel/

Mplus: Stand-alone Software

Mplus comes with a demo version, user's guide, and examples. It runs on Windows, Mac OS X, and Linux. The company provides short courses and web training sessions. Mplus uses a simple set of commands to specify and test many different types of SEM models. Mplus in R, for example, was specifically written to conduct Monte Carlo simulation and compare many alternative models. Some other features in Mplus are Bayesian SEM, exploratory SEM, power calculations, Item Response Theory (IRT), missing data routines, randomized trials, growth mixture models, and latent class analysis. Many examples, white papers, and references are provided for SEM researchers. More information about Mplus can be found at: www.statmodel.com/

Mx: Stand-alone Software (Free)

Mx is a matrix algebra and numeric optimizer for structural equation modeling that operates in Windows and Unix (Linux) operating systems. It comes with a user manual that covers matrix algebra, an Mx graphical user interface (MS Windows or MS-DOS only), example Mx script syntax, and examples. Some of the features in this software are the use of start values, selection of fit functions, power calculations, mixture modeling, non-linear data analysis, ordinal data analysis, and bootstrap estimates of confidence intervals. More information about Mx can be found at: www.vcu.edu/mx/

OpenMx: Stand-alone Software (Free); R Interface

OpenMx is free and open-source software for use with R that allows estimation of a wide variety of advanced multivariate statistical models. It consists of a library of functions and optimizers that allow you to quickly define an SEM model and estimate parameters given observed variables. It runs on Mac OS X, Windows (XP, Vista, 7, 8), and Linux operating systems. The same script programs in Windows will run in Mac OS X or Linux. OpenMx can be used by drawing path models or specifying models in terms of matrix algebra. It is extremely functional because it takes advantage of the R programming environment. The path model is specified and data set acquired using the R command language. More information about OpenMx can be found at: openmx.psyc.virginia.edu/

PROC CALIS (SAS)

PROC CALIS is a procedure in the SAS statistical package. It uses SAS data sets and specific SAS syntax code. Statisticians who frequently use SAS also use this program to conduct a test of their SEM model. Some of its features are the flexibility to describe a model with structural equation statements similar to EQS using the LINEQS option, or to specify complex matrix models using MATRIX statements. The PROC CALIS procedure also offers several optimization algorithms (NLOPTIONS) to obtain parameter estimates and standard errors when problems are encountered in the data due to non-normality and non-linearity. More information on this procedure can be found at: support.sas.com/documentation/cdl/en/statug/63033/HTML/default/viewer.htm#calis_toc.htm

R: Stand-alone Software (Free)

R is a library of packages with functions that permit flexible writing of script files to conduct statistical analysis in Windows, Mac OS X, or Linux operating systems. Two structural equation modeling packages available in R are *sem* and *lavaan*. Both packages can be installed using the *Load Package* command in the pull-down menu once the R software is installed. Alternatively, the command, install.packages("sem") or install.packages("lavaan") can be issued in the R console window of the software. The *sem* and *lavaan* packages require commands with arguments that specify the SEM model, parameter names, data file input, start values, optimization routine, and missing data. You can obtain more information and examples for either package by typing ?sem or ?lavaan in the R console window once the software is installed. Fox (2006) published an article about the *sem* package in the *Structural Equation Modeling* journal. Yves (2012) published an article about the *lavaan* package in the *Journal of Statistical Software*. A unique feature in the *lavaan* package is the *mimic* argument which permits similarity to either the Mplus (default) or EQS program. The *representation* argument permits similarity to the LISREL9 matrix output. More information about R, R installation, and R packages can be found at: www.cran.r-project.org

SEPATH (Statistica)

SEPATH is included in *Statistica* as a collection of procedures with data sets and examples. The SEM model is specified and the data set input into SEPATH via a dialog box in *Statistica*. Dialog boxes also permit the selection of analysis parameters (type of matrix to analyze, estimation method, constraints, etc.) and report the results (parameter estimates, standard errors, etc.). A path diagram of the SEM model can also be generated. Some features of this software are the

Wizards used to create CFA and SEM models; Monte Carlo analysis, and explanation of important concepts in SEM (iteration, estimation, model identification, non-centrality based fit indices, standard error differences between correlation versus covariance matrix input). You can find more information about SEPATH at: documentation.statsoft.com/STATISTICAHelp.aspx?path=SEPATH/Indices/SEPATHAnalysis_HIndex

SEM (STATA)

SEM is included in the *STATA* statistical package as a collection of procedures with data sets and examples. The company provides video tutorials, training sessions, and short courses. The software is available on the Windows, Mac OS X, and UNIX (Linux) operating systems. SEM uses either the *SEM Builder* option or command language to create path diagrams of models. *SEM Builder* uses the drag-n-drop approach to create the path diagrams, run the model, and display the results on the path diagram. Some features of interest are the mediation analysis, MIMIC models, latent growth curve models, multi-level models, and generalized linear models with binary, count, ordinal, and nominal outcomes. More information about SEM in *STATA* can be found at: www.stata.com/features/structural-equation-modeling/

Summary

We have listed many of the different SEM software programs available today for the SEM modeler and explained some of the features available in each. If you are new to structural equation modeling, check what software is available at your company or university. You may wish to learn the software program that comes with a statistics package (IBM—AMOS, SAS—PROC CALIS, STATA—SEM, Statistica—SEPATH). Alternatively, you can purchase a stand-alone software package (EQS, LISREL, Mplus) or download software that is free (Mx, R—sem, R—lavaan). Many researchers choose SEM software based on whether it runs in a Windows, Mac OS X, or UNIX (Linux) environment on their computer. We will demonstrate regression models, path models, confirmatory factor models, and the various SEM models in the book using some of these SEM software programs. We understand that the reader will not have access to most of these software programs; however, the data sets and examples in the book can be run with the SEM software available to you. The data and programs on the book website were saved as ASCII files, which permit easy copying and use in any SEM software program. The choice and installation of software to use with the data sets and examples is left up to the reader.

BOOK WEBSITE

This book has a website located at www.routledge.com/9781138811935/. From the website you will be able to access or download the following resources:

Computer files and/or diagrams for the SEM modeling types presented in the book, which include EQS, LISREL, Mplus, and R examples

Data sets used in the SEM model examples

Links to web pages that have more information about SEM

Suggested readings or journal articles in PDF format for many of the SEM topics covered in the book.

The purpose of the website is to support learning about how the different SEM models are created and tested using a variety of SEM software programs. It also allows for expansion of the website to include other data sets, examples, and software usage in the future. The availability of data and syntax files will assist in the analysis and interpretation of the examples in the book using the corresponding software program.

You will be able to open the syntax files from the different software programs in *Notepad* because they will be in ASCII text format. This provides a visual inspection of how each software program has unique syntax commands, but the specification of the model equations would be similar in nature. The basic equation structure, the output information, and the model diagram would all be similar. Some of the exercises involve real data sets, which will provide a practical guide for programming, analysis, and interpretation. Some of the SEM model examples in the book may only contain the solution and interpretation of results with the data and program on the website.

SUMMARY

In this chapter we briefly described structural equation modeling, presented the history of structural equation modeling, and explained four reasons why researchers would choose to conduct structural equation modeling. We introduced structural equation modeling by describing basic types of variables—that is, latent, observed, independent, and dependent—and basic types of SEM models—that is, regression, path, confirmatory factor, and structural equation models. The chapter concluded with a brief listing of the different structural equation modeling software programs and where to obtain them. The website contains data sets, programs, links, and other resources referenced in the book.

EXERCISES

1. Define the following terms:
   a. Latent variable
   b. Observed variable
   c. Dependent variable
   d. Independent variable
2. Explain the difference between a dependent latent variable and a dependent observed variable.
3. Explain the difference between an independent latent variable and an independent observed variable.
4. List the four reasons why a researcher would conduct structural equation modeling.

REFERENCES

Anderson, T. W., & Rubin, H. (1956). Statistical inference in factor analysis. In J. Neyman (Ed.), *Proceedings of the third Berkeley symposium on mathematical statistics and probability, Vol. V* (pp. 111–150). Berkeley: University of California Press.

Blalock, H. M., Jr. (1972). *Causal Models in the Social Sciences* (Editor). London: Macmillan.

Cudeck, R., Du Toit, S., & Sörbom, D. (2001) (Eds). *Structural equation modeling: Present and future. A Festschrift in honor of Karl Jöreskog*. Lincolnwood, IL: Scientific Software International.

Delucchi, M. (2006). The efficacy of collaborative learning groups in an undergraduate statistics course. *College Teaching*, 54, 244–248.

Duncan, O. D. (1966). Path analysis: Sociological examples. *The American Journal of Sociology*, 72(1), 1–16.

Fox, J. (2006) Structural equation modeling with the sem package in R. *Structural Equation Modeling*, 13, 465–486.

Goldberg, L. (1990). An alternative "description of personality": Big Five factor structure. *Journal of Personality and Social Psychology*, 59, 1216–1229.

Hershberger, S. L. (2003). The growth of structural equation modeling: 1994–2001. *Structural Equation Modeling*, 10(1), 35–46.

Howe, W. G. (1955). *Some contributions to factor analysis* (Report No. ORNL-1919). Oak Ridge National Laboratory, Oak Ridge, TN.

Jöreskog, K. G. (1963). *Statistical estimation in factor analysis: A new technique and its foundation*. Stockholm: Almqvist & Wiksell.

Jöreskog, K. G. (1969). A general approach to confirmatory maximum likelihood factor analysis. *Psychometrika*, 34, 183–202.

Jöreskog, K. G. (1973). A general method for estimating a linear structural equation system. In A. S. Goldberger & O. D. Duncan (Eds.), *Structural equation models in the social sciences* (pp. 85–112). New York: Seminar.

Jöreskog, K. G., & Sörbom, D. (2006). LISREL for Windows [Computer software]. Lincolnwood, IL: Scientific Software International.

Keesling, J. W. (1972). *Maximum likelihood approaches to causal flow analysis.* Unpublished doctoral dissertation. Chicago: University of Chicago.

Lawley, D. N. (1958). Estimation in factor analysis under various initial assumptions. *British Journal of Statistical Psychology*, 11, 1–12.

Parkerson, J. A., Lomax, R. G., Schiller, D. P., & Walberg, H. J. (1984). Exploring causal models of educational achievement. *Journal of Educational Psychology*, 76, 638–646.

Pearson, E. S. (1938). *Karl Pearson. An appreciation of some aspects of his life and work.* Cambridge: Cambridge University Press.

Spearman, C. (1904). The proof and measurement of association between two things. *American Journal of Psychology*, 15, 72–101.

Spearman, C. (1927). *The abilities of man.* New York: Macmillan.

Wiley, D. E. (1973). The identification problem for structural equation models with unmeasured variables. In A. S. Goldberger, & O. D. Duncan (Eds.), *Structural equation models in the social sciences* (pp. 69–83). New York: Seminar.

Wright, S. (1918). On the nature of size factors. *Genetics*, 3, 367–374.

Wright, S. (1921). Correlation and causation. *Journal of Agricultural Research*, 20, 557–585.

Wright, S. (1934). The method of path coefficients. *Annals of Mathematical Statistics*, 5, 161–215.

Wold, H. O. A. (1954). Causality and econometrics. *Econometrica*, 22(2), 162–177.

Yves, R. (2012). lavaan: An R package for structural equation modeling. *Journal of Statistical Software*, 48(2), 1–36.

Chapter 2

DATA ENTRY AND EDIT ISSUES

CHAPTER CONCEPTS

Data set formats
Data editing issues
  Measurement scale
  Restriction of range
  Missing data
  Outliers
  Linearity
  Non-normality
  Sample size and degrees of freedom
Start values
Non-positive definite matrices
Heywood case (negative variance)

DATA SET FORMATS

An important first step in using an SEM software program is to be able to enter raw data and/or import data. Many of the software programs permit options to read in text files (ASCII), comma separated files (CSV), data from other programs, such as SPSS, SAS, or EXCEL, enter correlation or covariance matrices directly into a program, or read in raw data and save as a system file.

A system file is a file type unique to the software program, for example, PRELIS files are system files in LISREL that appear in spreadsheet format. The spreadsheet format is similar to that found in SPSS, and permits analysis of statistics in the pull-down menu, and the saving of files that contain the variance–covariance matrix, the correlation matrix, means, and standard deviations of variables. The PRELIS system file activates a pull-down menu that permits data editing features, data transformations, statistical analysis of data, graphical display of data, multi-level modeling, and many other related features. Statistical packages that provide SEM software permit the researcher to conduct much needed data screening and editing (SPSS—AMOS, SAS—Proc Calis, Statistica—SEPATH, STATA—SEM). Nothing prevents a researcher from using a statistical package to edit data prior to using it in a stand-alone SEM software program (EQS, Mplus, Mx, R).

A researcher should also be aware of important data editing features needed prior to SEM analysis, whether available in a statistics package or the SEM software. This involves checking for missing data, conducting homogeneity of variance tests, normality tests, and data output options. An important data option is to save raw data in an ASCII tab delimited file or comma separated file. Many of the SEM software programs can read in the raw data in these file types. The data output options might also permit saving different types of variance–covariance matrices, descriptive statistics, or scores for future use. Your knowledge of data input and output features is very important, and tutorials or workshops are offered for many of the statistical packages and SEM software packages.

The statistical analysis of data in SEM requires you to become familiar with the software input and output data features. The most common approaches have been inputting raw data from ASCII data files (txt; dat), comma separated data files (csv), variance–covariance matrices, or correlation matrices with means and standard deviations. SEM software will convert raw data or the correlation matrix with means and standard deviations to a variance–covariance matrix in modeling applications. SEM software output varies, with some providing graphs, and others having options to output scores, different types of correlation matrices, or saving an asymptotic variance–covariance matrix.

DATA EDITING ISSUES

There are several key issues in the field of statistics that impact our analyses once data have been imported into a software program. These data issues are commonly referred to as the measurement scale of variables, restriction in the range of data, missing data values, outliers, linearity, and non-normality. Each of these data issues will be discussed because they not only affect traditional statistics, but present additional problems and concerns in structural equation modeling.

Measurement Scale

How variables are measured or scaled influences the type of statistical analyses we perform (Anderson, 1961; Stevens, 1946). Properties of scale also guide our understanding of permissible mathematical operations. For example, a nominal variable implies mutually exclusive groups; a biological gender has two mutually exclusive groups, male and female. An individual can only be in one of the groups that define the levels of the variable. In addition, it would **not** be meaningful to calculate a mean and a standard deviation on the variable *gender*. Consequently, the number or percentage of individuals at each level of the gender variable is the only mathematical property of scale that makes sense.

An ordinal variable, for example, attitude toward school, that is scaled *strongly agree*, *agree*, *neutral*, *disagree*, and *strongly disagree*, implies mutually exclusive categories that are ordered or ranked. When levels of a variable have properties of scale that involve mutually exclusive groups that are ordered, only certain mathematical operations are permitted. A comparison of ranks between groups is a permitted mathematical operation, while calculating a mean and standard deviation is not.

Final exam scores, an example of an interval variable, possess the property of scale that implies equal intervals between the data points, but no true zero point. This property of scale permits the mathematical operation of computing a mean and a standard deviation. Similarly, a ratio variable, for example, weight, has the property of scale that implies equal intervals and a true zero point (weightlessness). Therefore, ratio variables also permit mathematical operations of computing a mean and a standard deviation. The difference between interval and ratio data is whether a true zero point exists. For interval data, it is wise to have an arbitrary zero point whenever possible. This aids our interpretation in many statistical applications.

Our use of different variables requires us to be aware of their properties of scale and what mathematical operations are possible and meaningful, especially in SEM, where variance–covariance matrices are used. Different types of correlation coefficients will produce different types of variance–covariance matrices (Pearson, polychoric, polyserial) among variables depending upon the level of measurement, and this creates a unique problem in SEM. It is very important to understand how continuous variables, ordinal variables, and group or categorical variables are used in SEM. We will address the use of different variable types and correlation matrices in subsequent chapters of the book.

Restriction of Range

Data values at the interval or ratio level of measurement can be further defined as being discrete or continuous. For example, final exam scores could be reported in whole numbers (discrete). Similarly, the number of children in a family would be considered a discrete level of measurement, that is, 5 children. In contrast, a continuous variable is reported using decimal values, for example, a student's grade point average would be reported as 3.75 on a 5-point scale.

Karl Jöreskog and Dag Sörbom (1996) provided a criterion based on their research that defines whether a variable is ordinal or interval. If a variable has fewer than 15 scale points, it should be considered an ordinal variable, whereas a variable with 15 or more scale points is considered a continuous variable. This 15 scale point criterion permits the Pearson correlation coefficient values to vary between +/– 1.0. Variables with fewer distinct scale points restrict the value of the Pearson correlation coefficient such that it may only vary between +/– 0.5. SEM software requires you to declare the type of variables being used in the model. Other factors that affect the Pearson correlation coefficient are presented in this chapter.

Missing Data

The statistical analysis of data is affected by missing data values in variables. That is, not every subject has an actual value for every variable in the data set, as some values are missing. It is common practice in statistical packages to have default values for handling missing values. Caution is needed because SEM software varies on what missing data values are permitted. Also, an ASCII tab-delimited data file with blanks for missing data can cause data input errors. The researcher has the options of deleting subjects who have missing values, replacing the missing data values, or using robust statistical procedures that accommodate for the presence of missing data.

The various SEM software programs handle missing data differently and have different options for replacing missing data values. Table 2.1 lists many of the various options for dealing with missing data. These options can dramatically affect the number of subjects available for analysis, the magnitude and direction of the correlation coefficient, or create problems if means, standard deviations, and correlations are computed based on different sample sizes.

The *Listwise* deletion of cases and *Pairwise* deletion of cases are not always recommended options due to the possibility of losing a large number of subjects, thus dramatically reducing the sample size, and affecting parameter estimates and standard errors. *Mean substitution* works best when only a small number of missing values are present in the data, whereas *regression imputation* provides a useful approach with a moderate amount of missing data.

Schafer (1997) pointed out three types of non-response for missing values: MCAR (missing completely at random), MAR (missing at random), and MNAR (missing

Table 2.1: Options for Dealing with Missing Data

Listwise	Delete subjects with missing data on any variable
Pairwise	Delete subjects with missing data on each pair of variables used
Mean substitution	Substitute the mean for missing values of a variable
Regression imputation	Substitute a predicted value for the missing value of a variable
Expectation maximization (EM)	Find expected value based on expectation maximization algorithm
Matching response pattern	Match cases with incomplete data to cases with complete data to determine a missing value

not at random). MCAR does not depend on any variable in the data for the missingness. MAR usually implies that variable missingness is related to other variables. MNAR does depend on response bias for the missingness, for example, males and females respond differently. The expectation maximization (EM), Monte Carlo Markov Chain (MCMC), and matching response pattern approaches have been recommended when larger amounts of data are missing at random. The EM algorithm generates one single complete data matrix whereas the MCMC method generates several complete data matrices and uses the average. The MCMC method is more reliable than the EM algorithm. In both cases, the complete data matrix can be used to estimate the mean vector and the covariance matrix of the observed variables which can be used in SEM to estimate parameters in the model.

More information about how to handle missing data using MCMC with Mplus programs is available in Enders (2013), while others address general missing data issues: McKnight, McKnight, Sidani and Aurelio (2007), and Peng, Harwell, Liou, and Ehman (2007). Davey and Savla (2009) have recently published an excellent book with SAS, SPSS, STATA, and Mplus source programs to handle missing data in SEM, especially in the context of power analysis.

LISREL Example
The imputation of missing values in LISREL9 uses the *matching response pattern* approach with raw data in a system file. The matching response pattern approach substitutes the missing value of a single case on a variable with a value based on cases of other variables that have complete data, thus using the response pattern over the set of variables with complete data. Multiple imputation is possible when missing values occur on more than one variable by selecting mean substitution, the expectation maximization algorithm (EM) or the Monte Carlo Markov Chain (MCMC) option, which generates random draws from probability distributions via Markov chains. Once raw data are saved as a system file (*.lsf), a main menu appears with a pull-down menu. The *Statistics* menu has the option of *Impute Missing Value* (response matching approach) or the *Multiple Imputation* (mean substitution, EM, MCMC).

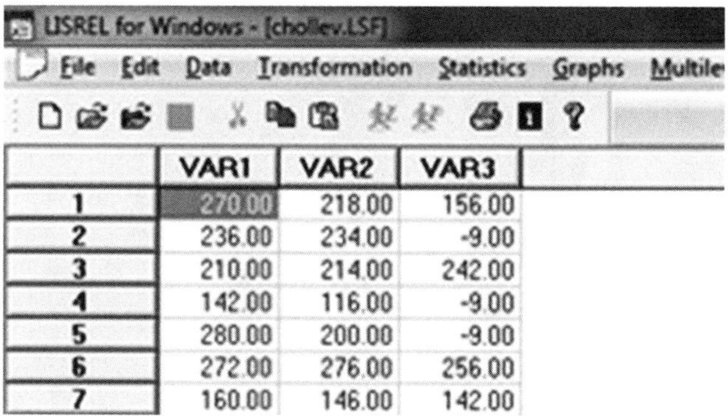
LISREL for Windows - [chollev.LSF]
File Edit Data Transformation Statistics Graphs Multile
VAR1 VAR2 VAR3
1 270.00 218.00 156.00
2 236.00 234.00 -9.00
3 210.00 214.00 242.00
4 142.00 116.00 -9.00
5 280.00 200.00 -9.00
6 272.00 276.00 256.00
7 160.00 146.00 142.00

■ **Figure 2.1:** LISREL MAIN MENU SCREEN

We present an example using data on the cholesterol levels of 28 patients treated for heart attacks. We assume the data to be missing at random (MAR) with an underlying multivariate normal distribution. Cholesterol levels were measured after 2 days (VAR1), after 4 days (VAR2), and after 14 days (VAR3), but only 19 of the 28 patients had complete data. The missing value was entered as –9.000. The raw data are from the file, *chollev.dat*, located in the Tutorial subfolder on the computer directory, LISREL9 Student Examples. We used the import data feature, which prompted us to save the raw data file as a LISREL system file type (*chollev.LSF*). The LISREL main menu should look like the image in Figure 2.1.

We next click on Statistics on the tool bar menu and select Impute Missing Values from the pull-down menu. This provides a dialog box with the variable that has missing values (VAR3), and the variables with complete cases to be used as matching variables (VAR1 and VAR2). The dialog box is shown in Figure 2.2.

We next selected Output Options and saved the transformed data in a new system file, *cholnew.lsf*, and output the new correlation matrix (*chol.cor*), means (*chol.mn*), and standard deviations (*chol.sd*) into saved files. These files were saved on the book website. The Output dialog box is shown in Figure 2.3.

After clicking on Run, a computer output file appears with the results. *Do not be upset about the warning messages.* They are indicating that the variables are continuous, which is defined by having more than 15 categories. Also, notice that the output indicates no missing values after imputation, as shown in Table 2.2.

We should examine our data both before (Table 2.3) and after (Table 2.4) imputation of missing values. Here, we used the matching response pattern method. This comparison provides us with valuable information about the nature of the missing data.

■ **Figure 2.2:**   IMPUTE MISSING VALUES DIALOG BOX

■ **Figure 2.3:**   OUTPUT DIALOG BOX

Table 2.2: LISREL Imputation Output

W_A_R_N_I_N_G: VAR1 has more than 15 categories and will be treated as continuous. ERROR CODE 201.
W_A_R_N_I_N_G: VAR2 has more than 15 categories and will be treated as continuous. ERROR CODE 201.
W_A_R_N_I_N_G: VAR3 has more than 15 categories and will be treated as continuous. ERROR CODE 201.
Number of Missing Values per Variable After Imputation

VAR1	VAR2	VAR3
0	0	0

Table 2.3: Summary Statistics with Missing Data

Total Effective Sample Size =   19
Number of Missing Values per Variable

VAR1	VAR2	VAR3
0	0	9

Univariate Summary Statistics for Continuous Variables

Variable	Mean	St. Dev.	Skewness	Kurtosis	Minimum	Freq.	Maximum	Freq.
VAR1	253.929	47.710	−0.351	0.473	142.000	1	360.000	1
VAR2	230.643	46.967	0.019	1.344	116.000	1	352.000	1
VAR3	221.474	43.184	−0.166	−0.890	142.000	1	294.000	1

Correlation Matrix

	VAR1	VAR2	VAR3
VAR1	1.000		
VAR2	0.689	1.000	
VAR3	0.393	0.712	1.000

The correlation, mean, and standard deviation of VAR3 is mildly affected by the missing data. The imputation of missing values for VAR3 indicated a small reduction in mean value (221.47 vs 220.71), a small reduction in the standard deviation (43.18 vs 42.77), and a slight change in the bivariate correlation coefficients. The concern is whether to proceed with the imputed values or drop subjects with missing values, thus reducing the sample size to $n = 19$. We do not want correlations based on different sample sizes, so an important decision must be made. Because the imputed values only mildly changed the descriptive statistics, the parameter estimates from a statistical analysis would probably not be substantially different. However, it would be prudent to make a comparison of results before and after imputation to determine how much the parameter estimates and standard errors are affected by missing data.

Another option is to use the new feature in LISREL9, namely, *full information maximum likelihood estimation*, which is *automatically* activated when using a raw

Table 2.4: Summary Statistics with Imputation

Total Sample Size (N) =   28
Number of Missing Values per Variable After Imputation

VAR1	VAR2	VAR3
--------	--------	--------
0	0	0

Imputations for   VAR3

Case	2	imputed with value	204	(Variance Ratio = 0.000), NM =	1	
Case	4	imputed with value	142	(Variance Ratio = 0.000), NM =	1	
Case	5	imputed with value	182	(Variance Ratio = 0.000), NM =	1	
Case	10	imputed with value	280	(Variance Ratio = 0.000), NM =	1	
Case	13	imputed with value	248	(Variance Ratio = 0.000), NM =	1	
Case	16	imputed with value	256	(Variance Ratio = 0.000), NM =	1	
Case	18	imputed with value	216	(Variance Ratio = 0.000), NM =	1	
Case	23	imputed with value	188	(Variance Ratio = 0.000), NM =	1	
Case	25	imputed with value	256	(Variance Ratio = 0.000), NM =	1	

Univariate Summary Statistics for Continuous Variables

Variable	Mean	St. Dev.	Skewness	Kurtosis	Minimum	Freq.	Maximum	Freq.
VAR1	253.929	47.710	−0.351	0.473	142.000	1	360.000	1
VAR2	230.643	46.967	0.019	1.344	116.000	1	352.000	1
VAR3	220.714	42.771	−0.208	−0.910	142.000	2	294.000	1

Correlation Matrix

	VAR1	VAR2	VAR3
	--------	--------	--------
VAR1	1.000		
VAR2	0.673	1.000	
VAR3	0.404	0.787	1.000

data file with missing data. The term *full information maximum likelihood* (FIML) is used with data that have missing values, while the term *maximum likelihood* (ML) is used with complete data. In LISREL9, however, you can still perform the steps outlined above to gain knowledge about the nature of the missingness in the data.

We highly recommend comparing any analyses before and after the replacement of missing data values to fully understand the impact missing data values have on the parameter estimates and standard errors. Given the high-speed computers and software of today, we also recommend checking how the replacement of missing values using the EM, MCMC, and FIML approaches affect your statistical analysis. It would be easy to compare the different ways missing data affect parameter estimates and standard errors in a model, that is, analysis with missing data, analysis with imputed data, and analysis with automatic FIML. A comparison of these methods is also warranted in multiple variable imputations to determine the effect of using a different algorithm on the replacement of missing values.

Outliers

Outliers or influential data points can be defined as data values that are extreme or atypical on either the independent (X) or dependent (Y) variables or both (Ho & Naugher, 2000). Outliers can occur as a result of observation errors, data entry errors, instrument errors based on layout or instructions, or actual extreme values from self-report data. Because outliers affect the mean, the standard deviation, and correlation coefficient values, they must be explained, deleted, or accommodated by using robust statistics. Sometimes, additional data will need to be collected to **fill in** the gap along either the Y or X axes. Statistical packages today have outlier detection methods available that include the following: box plot display, scatterplot, histogram, and frequency distributions.

Linearity

Some statistical calculations, such as Pearson correlation, assume that the variables are linearly related to one another. Thus, a standard practice is to visualize the coordinate pairs of data points of two continuous variables by plotting the data in a scatterplot. These bivariate plots depict whether the data are linearly increasing or decreasing. The presence of curvilinear data reduces the magnitude of the Pearson correlation coefficient, even resulting in the presence of a zero correlation. Recall that the Pearson correlation value indicates the magnitude and direction of the *linear* relationships between two variables. Figure 2.4 shows the importance of visually displaying the bivariate data scatterplot.

Non-normality

In statistics, several transformations are given to handle issues with non-normal data. Some of these common transformations are described in Table 2.5.

Inferential statistics assumes data are normally distributed. Data that are *skewed* (lack of symmetry) frequently occur more often along one part of the measurement scale, and affect the variance–covariance among variables. In addition, *kurtosis* (peakedness) in data will impact statistics. *Leptokurtic* data values are more peaked than the normal distribution, whereas *platykurtic* data values are flatter and more dispersed along the X axis, but have a consistent low frequency on the Y axis, that is, the frequency distribution of the data appears more rectangular in shape. One or more of the data transformation methods in Table 2.5 may correct

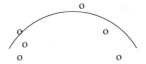

▪Figure 2.4: SCATTER PLOTS (LEFT: CORRELATION IS LINEAR. RIGHT: CORRELATION IS NON-LINEAR)

Table 2.5: Data Transformation Types

$y = \ln(x)$ or $y = \log_{10}(x)$ or $y = \ln(x + 0.5)$	Useful with clustered data or cases where the standard deviation increases with the mean
$y = \text{sqrt}(x)$	Useful with Poisson counts
$y = \arcsin((x + 0.375)/(n + 0.75))$	Useful with binomial proportions $[0.2 < p = x/n < 0.8]$
$y = 1/x$	Useful with gamma-distributed x variable
$y = \text{logit}(x) = \ln(x/(1 - x))$	Useful with binomial proportions $x = p$
$y = \text{normit}(x)$	Quantile of normal distribution for standardized x
$y = \text{probit}(x) = 5 + \text{normit}(x)$	Most useful to resolve non-normality of data

Note: probit(x) is same as normit(x) plus 5 to avoid negative values.

for skewness or leptokurtic values in the data; however, platykurtic data usually require recoding into categories or obtaining better data.

Non-normal data can occur because of the scaling of variables (ordinal rather than interval) or the limited sampling of subjects. Possible solutions for skewness are to resample more participants or to perform a linear transformation as indicated in Table 2.5. Our experience is that a probit data transformation works best in correcting skewness. Kurtosis in data is more difficult to resolve; some possible solutions include additional sampling of subjects, or the use of bootstrap methods, normalizing scores, or alternative methods of parameter estimation, for example, weighted least squares (WLS).

The presence of skewness and kurtosis can be detected by running descriptive statistics in many of the available software packages. Many of the SEM software programs routinely output this information. The output also includes both univariate and multivariate tests of normality. In SEM, the handling of non-normal data is typically done by a data transformation algorithm (Table 2.5), or using an asymptotic covariance matrix as input along with the sample covariance matrix. In LISREL9, this is done by the commands:

```
Covariance matrix from file boy.cov
Asymptotic covariance matrix from file boy.acm
```

We can use the asymptotic covariance matrix in two different ways: (a) as a weight matrix when specifying the method of estimation as weighted least squares (WLS), and (b) as a weight matrix that adjusts the normal-theory weight matrix to correct for bias in standard errors and model-fit statistics.

We generally consider normality of continuous variables with a variance–covariance matrix. Researchers should not mix correlation types; Pearson correlation with continuous variables is different than tetrachoric correlations with ordinal variables. A researcher should understand the variables level of measurement, type of correlation, and whether normality is an issue. If skewed non-normal data are present for a continuous variable, then consider a linear transformation or use of an **asymptotic variance–covariance matrix.**

Structural equation modeling (SEM) was introduced initially as a way of analyzing a covariance or correlation matrix. Typically, one would read this matrix into an SEM program and estimate the model parameter estimates using maximum likelihood estimation. If raw data were available without missing values, one would first estimate an asymptotic covariance matrix to obtain robust estimates of standard errors and chi-squares. Modern structural equation modeling programs use raw data, variance–covariance, or correlation matrices.

There are advantages to using a raw data file, namely, handling missing data values and non-normality issues. For example, in LISREL9 (Jöreskog, Sörbom, du Toit, & du Toit, 2001), the analysis will automatically perform robust estimation of standard errors and chi-square goodness-of-fit measures under non-normality. It is no longer necessary to estimate an asymptotic covariance matrix and read this into a program with the sample covariance matrix. The estimation of the asymptotic covariance matrix is now done in LISREL9, and if data contain missing values, the software will automatically use full information maximum likelihood (FIML) to estimate the model parameters. Previously, a researcher would impute missing values by EM or MCMC and estimate model parameters based on the imputed data. Today, this works for both continuous and ordinal variables. EQS software (Bentler & Wu, 2002) also has elliptical estimation methods that can be used in the presence of non-normal data distributions.

SUMMARY

Structural equation modeling is a correlation research method; therefore, the measurement scale, restriction of range in the data values, missing data, outliers, non-linearity, and non-normality of data affect the variance–covariance among variables and thus can impact the SEM analysis. Researchers should use the built-in menu options in a statistics package or SEM program to examine, graph, and test for any of these problems in the data prior to conducting any SEM model analysis. Basically, researchers should know their data characteristics. Data screening is a very important first step in structural equation modeling. In the next chapter, we provide specific examples to illustrate the importance of the topics covered in this chapter.

Table 2.6 lists the data editing issues along with suggestions for correcting any of these issues in the data.

Table 2.6: Data Editing Issues with Suggestions

Issue	Suggestions
Measurement scale	Need to take the measurement scale of the variables into account when computing statistics such as means, standard deviations, and correlations.
Restriction of range	Need to consider range of values obtained for variables, as restricted range of one or more variables can reduce the magnitude of correlations.
Missing data	Need to consider missing data on one or more subjects for one or more variables as this can affect SEM results. Cases are lost with listwise deletion, pairwise deletion is often problematic (for example, different sample sizes), and thus modern imputation methods are recommended.
Outliers	Need to consider outliers as they can affect statistics such as means, standard deviations, and correlations. They can either be explained, deleted, or accommodated (using either robust statistics or obtaining additional data to fill-in). Can be detected by methods such as box plots, scatterplots, histograms or frequency distributions.
Linearity	Need to consider whether variables are linearly related, as non-linearity can reduce the magnitude of correlations. Can be detected by scatter plots. Can be dealt with by transformations or deleting outliers.
Non-normality	Need to consider whether the variables are normally distributed, as non-normality can affect resulting SEM statistics. Can be detected by univariate tests, multivariate tests, and skewness and kurtosis statistics. Can be dealt with by transformations, additional sampling, bootstrapping, normalizing scores, or alternative methods of estimation.

EXERCISES

1. Define the following levels of measurement:
   a. Nominal
   b. Ordinal
   c. Interval
   d. Ratio
2. Describe the different imputation methods.
3. Explain how each of the following affects statistics:

a. Restriction of range
b. Missing data
c. Outliers
d. Non-linearity
e. Non-normality

REFERENCES

Anderson, N. H. (1961). Scales and statistics: Parametric and non-parametric. *Psychological Bulletin*, 58, 305–316.

Bentler, P., & Wu, E. J. C. (2002). EQS user's guide. Encino, CA: Multivariate Software, Inc.

Davey, A., & Savla, J. (2009). *Statistical power analysis with missing data: A structural equation modeling approach*. New York: Routledge, Taylor & Francis Group.

Enders, C. K. (2013). Analyzing structural equation models with missing data. In G. R. Hancock & R. O. Mueller (Eds.), *Structural equation modeling: A second course* (2nd edn.). Greenwich, CT: Information Age, pp. 493–519.

Ho, K., & Naugher, J. R. (2000). Outliers lie: An illustrative example of identifying outliers and applying robust methods. *Multiple Linear Regression Viewpoints*, 26(2), 2–6.

Jöreskog, K. G., & Sörbom, D. (1996). *PRELIS2: User's reference guide*. Lincolnwood, IL: Scientific Software International.

Jöreskog, K., Sörbom, D., du Toit, S., & du Toit, M. (2001). *LISREL8: New statistical features*. Chicago, IL: Scientific Software International.

McKnight, P. E., McKnight, K. M., Sidani, S., & Aurelio, J. F. (2007). *Missing data: A gentle introduction*. New York: Guilford.

Peng, C.-Y. J., Harwell, M., Liou, S.-M., & Ehman, L. H. (2007). Advances in missing data methods and implications for educational research. In S. S. Sawilowsky (Ed.), *Real data analysis*. Charlotte: Information Age.

Schafer, J. L. (1997). *Analysis of incomplete data*. Boca Raton, FL: Chapman & Hall/CRC Press.

Stevens, S. S. (1946). On the theory of scales of measurement. *Science*, 103, 677–680.

Chapter 3

CORRELATION

CHAPTER CONCEPTS

Types of correlation coefficients
Factors affecting correlation
Outliers
Correction for attenuation
Non-positive definite matrices
Bivariate, part, and partial correlation
What is a suppressor variable?
Correlation versus causation

In the previous chapter, we explained several data preparation issues that can affect structural equation modeling. In this chapter, we further explain how data issues can affect correlation, and hence variance–covariance matrices used in SEM. This discussion centers around a number of factors that affect correlation coefficients, as well as the assumptions and limitations of correlation methods in structural equation modeling.

TYPES OF CORRELATION COEFFICIENTS

Sir Francis Galton conceptualized the correlation and regression procedure for examining covariance in two or more traits, and Karl Pearson (1896) developed the statistical formula for the correlation coefficient and regression based on his suggestion (Crocker & Algina, 1986; Ferguson & Takane, 1989; Tankard, 1984). Shortly thereafter, Charles Spearman (1904) used the correlation procedure to develop a factor analysis technique. The correlation, regression, and factor analysis techniques have for many decades formed the basis for generating tests and defining constructs. Today, researchers are expanding their understanding of the roles that correlation, regression, and factor analysis play in theory and construct definition to include latent variable, covariance structure, and confirmatory factor measurement models.

The relationships and contributions of Galton, Pearson, and Spearman to the field of statistics, especially correlation, regression, and factor analysis, are quite interesting (Tankard, 1984). In fact, the basis of association between two variables, that is bivariate correlation, has played a major role in statistics. The Pearson correlation coefficient provides the basis for point estimation (test of significance), explanation (variance accounted for in a dependent variable by an independent variable), prediction (of a dependent variable from an independent variable through linear regression), reliability estimates (test–retest, equivalence), and validity (factorial, predictive, concurrent).

The Pearson correlation coefficient also provides the basis for establishing and testing models among measured and/or latent variables. The partial and part correlations further permit the identification of specific bivariate relations between variables that allow for the specification of unique variance shared between two variables while controlling for the influence of other variables. Partial and part correlations can be tested for significance, similar to the Pearson correlation coefficient, by simply using the degrees of freedom, $n - 2$, in the standard correlation table of significance values (Table A.3) or an F test in multiple regression which tests the difference in R^2 values between full and restricted models (Tables A.5 and A.6).

Although the Pearson correlation coefficient has had a major impact in the field of statistics, other correlation coefficients have emerged depending upon the level of variable measurement. Stevens (1968) provided the properties of scales of measurement that have become known as nominal, ordinal, interval, and ratio. The types of correlation coefficients developed for these various levels of measurement are categorized in Table 3.1.

Many popular computer programs, for example, SAS and SPSS, typically do not compute all of these correlation types. Therefore, you may need to check a popular statistics book or look around for a computer program (R software) that will compute the type of correlation coefficient you need, for example, the phi and point-biserial coefficient are usually computed using a measurement software package. The Pearson coefficient, tetrachoric or polychoric (for several ordinal variable pairs) coefficient, and biserial or polyserial (for several continuous and ordinal variable pairs) coefficient are typically used in SEM software (AMOS, EQS, LISREL, Mplus, etc.). Although SEM software programs include different correlations for different types of models, the use of variables with different levels of measurement has traditionally been a problem in the field of statistics, for example, in multiple regression and multivariate statistics.

Table 3.1: Types of Correlation Coefficients

Correlation Coefficient	Level of Measurement
Pearson product-moment	Both variables interval
Spearman rank, Kendall's tau	Both variables ordinal
Phi, contingency	Both variables nominal
Point biserial	One variable interval, one variable dichotomous
Gamma, rank biserial	One variable ordinal, one variable nominal
Biserial	One variable interval, one variable artificial[a]
Polyserial	One variable interval, one variable ordinal with underlying continuity
Tetrachoric	Both variables dichotomous (nominal artificial[a])
Polychoric	Both variables ordinal with underlying continuities

[a] *Artificial* refers to recoding variable values into a dichotomy.

FACTORS AFFECTING CORRELATION COEFFICIENTS

Given the important role that correlation plays in structural equation modeling, we need to understand the factors that affect establishing relations among multivariable data points. The key factors are the level of measurement, restriction of range in data values (variability, skewness, kurtosis), missing data, non-linearity, outliers, correction for attenuation, and issues related to sampling variation, confidence intervals, effect size, significance, sample size, and power. Some of these factors are illustrated in Table 3.2, where the complete data set indicates a Pearson correlation coefficient of $r = .782$, $p = .007$, for $n = 10$ pairs of scores. The non-linear data result in $r = 0$, the missing data an $r = .659$, $p = .10$ (non-significant), restriction of range an $r = 0$, and sampling effect an $r = -1.0$, $p < .0001$ (a complete reversal of sign and direction). We see in Table 3.2 the dramatic impact these factors have on the Pearson correlation coefficient.

Non-linearity

The Pearson correlation coefficient indicates the degree of linear relation between two variables. It is possible that two variables can indicate no correlation if they have a curvilinear relation. Thus, the extent to which the variables deviate from the assumption of a linear relation will affect the size of the correlation coefficient. In Table 3.2, the non-linear data indicate a correlation of $r = 0.0$, which indicates no relation when the data are actually non-linear.

It is therefore important to check for linearity of the scores; the common method is to graph the coordinate data points in a scatterplot. The linearity assumption should not be confused with recent advances in testing interaction in structural equation models. You should also be familiar with the *eta* coefficient as an index

Table 3.2: Heuristic Data Sets

Complete Data (r = .782, p = .007)		Non-linear Data (r = 0.0, p = 1.0)		Missing Data (r = .659, p = .108)	
Y	X	Y	X	Y	X
8.00	6.00	1.00	1.00	8.00	—
7.00	5.00	2.00	2.00	7.00	5.00
8.00	4.00	3.00	3.00	8.00	—
5.00	2.00	4.00	4.00	5.00	2.00
4.00	3.00	5.00	5.00	4.00	3.00
5.00	2.00	6.00	5.00	5.00	2.00
3.00	3.00	7.00	4.00	3.00	3.00
5.00	4.00	8.00	3.00	5.00	—
3.00	1.00	9.00	2.00	3.00	1.00
2.00	2.00	10.00	1.00	2.00	2.00

Range of Data (r = 0.0, p = 1.0)		Sampling Effect (r = −1.0, p < .0001)	
Y	X	Y	X
3.00	1.00	8.00	3.00
3.00	2.00	9.00	2.00
4.00	3.00	10.00	1.00
4.00	4.00		
5.00	1.00		
5.00	2.00		
6.00	3.00		
6.00	4.00		
7.00	1.00		
7.00	2.00		

of non-linear relationship between two variables and with the testing of linear, quadratic, or cubic effects, which are covered in a multiple regression course.

Missing Data

If missing data were present, the Pearson correlation coefficient would drop to $r = .659$, $p = .108$, for $n = 7$ pairs of scores. The Pearson correlation coefficient changes from being statistically significant to not being statistically significant. More importantly, in a correlation matrix with several variables, the various correlation coefficients could be computed on different sample sizes. If we used *listwise deletion* of cases, then any variable in the data set with a missing value would cause a subject to be deleted, possibly causing a substantial reduction in our sample size, whereas *pairwise deletion* of cases would result in different sample sizes for our correlation coefficients in the correlation matrix.

Researchers have examined various aspects of how to handle or treat missing data beyond our introductory example using a small heuristic data set. One basic approach is to eliminate any observations where some of the data are missing, *listwise deletion*. Listwise deletion is not recommended because of the loss of information on other variables, and statistical estimates are based on reduced sample size. *Pairwise deletion* excludes data only when they are missing on the pairs of variables selected for analysis. However, this could lead to different sample sizes for the different correlations and related statistical estimates. A third approach, *data imputation*, replaces missing values with an estimate, for example, the mean value on a variable for all subjects who did not report any data for that variable (Beale & Little, 1975).

Missing data can arise in different ways (Little & Rubin, 1987, 1990; Schafer, 1997). *Missing completely at random* (MCAR) implies that data on variable X are missing unrelated statistically to the values that have been observed for other variables as well as X. *Missing at random* (MAR) implies that data values on variable X are missing conditional on other variables, but are unrelated to the values of X. A third situation (MNAR), *non-ignorable* data, has missing values because of a biased non-response, for example, males do not respond. In MNAR, missing values are input based on probabilistic information about the values that would have been observed. For MCAR data, mean substitution yields biased variance and covariance estimates, whereas listwise and pairwise deletion methods yield consistent solutions. For MAR data, mean substitution, listwise, and pairwise deletion methods produce biased results. When missing data are non-ignorable, all approaches yield biased results. It would be prudent for the researcher to investigate how parameter estimates and standard errors are affected by the use or non-use of a data imputation method. A few references are provided to give a more detailed understanding of missing data (Arbuckle, 1996; Davey & Savla, 2009; Enders, 2013; McKnight, McKnight, Sidani & Aurelio, 2007; Peng, Harwell, Liou & Ehman, 2007; Wothke, 2000).

Restriction of Range

Four types or levels of measurement typically define whether the scale interpretation of a variable is nominal, ordinal, interval, or ratio (Stevens, 1968). Initially, SEM required variables were measured at the interval or ratio level of measurement, so the Pearson product-moment correlation coefficient was used in regression, path, factor, and structural equation modeling. The interval or ratio scaled variable values should have a sufficient range of score values to introduce variance. If the range of scores is restricted, the magnitude of the correlation value is decreased. In Table 3.2, we see that a restriction of range changed the correlation from $r = .782$ to a value of $r = 0.0$.

If the distributions of the variables are widely divergent, correlation can also be affected, and so several data transformations were suggested by Ferguson and Takane (1989) to provide a closer approximation to a normal distribution in the presence of skewed or kurtotic data. Some possible transformations are the square root transformation (sqrt X), the logarithmic transformation (log X), the reciprocal transformation ($1/X$), and the arcsine transformation (arcsin X). The probit transformation appears to be most effective in handling univariate skewed data, but it is wise to check on which transformation works best.

Consequently, the level of variable measurement and the range of values for the measured variables can have profound effects on your statistical analysis (in particular, on the mean, variance, and correlation). The scale and range of a variable's numerical values affects statistical methods, and this is no different in structural equation modeling. Most SEM software today provides tests of normality, skewness, and kurtosis on variables, and computes an appropriate covariance matrix.

The meaningfulness of a bivariate correlation will also depend on the variables employed; hence, your theoretical perspective is very important. You may recall from your basic statistics course that a spurious correlation is possible when two sets of scores correlate significantly, but their relationship is not meaningful or substantive in nature.

OUTLIERS

The Pearson correlation coefficient can be drastically affected by a single outlier on X or Y, *or both*. For example, the two data sets in Table 3.3 indicate a $Y = 27$ value (Set A) versus a $Y = 2$ value (Set B) for the last subject. In the first set of data, $r = .524$, $p = .37$, whereas in the second set of data, $r = -.994$, $p = .001$. Is the $Y = 27$ data value an outlier based on limited sampling or is it

Table 3.3: Outlier Data Sets

Set A ($r = .524$, $p = .37$)		Set B ($r = -.994$, $p = .001$)	
X	Y	X	Y
1	9	1	9
2	7	2	7
3	5	3	5
4	3	4	3
5	27	5	2

Start Values

Regression, path, factor, and structural equation models mathematically solve a set of simultaneous equations typically using ordinary least squares (OLS) estimates as initial estimates of coefficients in the model. However, these initial estimates or coefficients are sometimes distorted or too different from the final admissible solution due to *messy* data. When this happens, more reasonable *start values* need to be chosen to permit convergence to a solution. It is easy to see from the basic regression coefficient formula that the correlation coefficient value and the standard deviation values of the two variables affect the initial OLS start value estimates:

$$b = r_{xy} \left(\frac{s_y}{s_x} \right).$$

Simultaneous equations are generally solved by initially setting parameter estimates at zero or using two-stage least squares estimates (2SLS), then several iterations using maximum likelihood estimation are conducted. When convergence to a solution does not occur, that is, parameter estimates, standard errors, and error covariance are not computed correctly, then start values can be attempted. EQS and LISREL permit user-specified start values, while Mplus permits user-specified or automatic start values.

Sample Size

A common formula used to determine sample size when estimating means of variables was given by McCall (1982): $n = (Z \, \sigma / \varepsilon)^2$, where n is the sample size needed for the desired level of precision, ε is the effect size, Z is the confidence level, and σ is the population standard deviation of scores (σ can be estimated from prior research studies, test norms, or the range of scores divided by 6). For example, given a random sample of American College Testing (ACT) scores from a defined population with a standard deviation of 100, a desired confidence level of 1.96 (which corresponds to a .05 level of significance), and an effect size of 20 (difference between sampled ACT mean and population ACT mean), the sample size needed would be $[1.96 \, (100)/20)]^2 = 96$.

In structural equation modeling, however, the researcher often requires a much larger sample size to maintain power and obtain stable parameter estimates and standard errors. The need for larger sample sizes is also due in part to the program requirements and the multiple observed variables used to define latent variables. Hoelter (1983) proposed the *critical N* statistic, which indicates the sample size needed to obtain a chi-square value that would reject the null hypothesis in a structural equation model. The required sample size and power estimates that provide a reasonable indication of whether a researcher's data fits their theoretical model or to estimate parameters is more difficult to specify in SEM.

SEM software programs estimate coefficients based on the user-specified theoretical model, or *implied model*, but also must work with the saturated and independence models. A *saturated model* is the model with all parameters indicated, while the *independence model* is the null model or model with no parameters estimated. A saturated model with p observed variables has $p(p + 3)/2$ free parameters. [*Note:* Number of independent elements in the symmetric covariance matrix = $p(p + 1)/2$. Number of means = p, so total number of independent elements = $p(p + 1)/2 + p = p(p + 3)/2$.] For example, with 10 observed variables, $10(10 + 3)/2 = 65$ free parameters. If the sample size is small, then there is not enough information to estimate parameters in the saturated model for a large number of variables. Consequently, the chi-square fit statistic and derived statistics such as Akaike's Information Criterion (AIC) and the root-mean-square error of approximation (RMSEA) cannot be computed. In addition, the fit of the independence model is required to calculate other fit indices such as the Comparative Fit Index (CFI) and the Normed Fit Index (NFI).

Ding, Velicer, and Harlow (1995) located numerous studies (Anderson & Gerbing, 1988) that were in agreement that 100 to 150 subjects is the *minimum* satisfactory sample size when conducting structural equation models. Boomsma (1982, 1983) recommended 400, and Hu, Bentler, and Kano (1992) indicated that in some cases 5000 is insufficient! Many of us may recall rules of thumb in our statistics texts, for example, 10 subjects per variable or 20 subjects per variable. Costello and Osborne (2005) demonstrated in their Monte Carlo study that 20 subjects per variable is recommended for best practices in factor analysis. In our examination of published SEM research, we have found that many articles used from 250 to 500 subjects, although the greater the sample size, the more likely it is one can validate the model using cross-validation methods. For example, Bentler and Chou (1987) suggested that a ratio as low as five subjects per variable would be sufficient for normal and elliptical distributions when the latent variables have multiple indicators and that a ratio of at least 10 subjects per variable would be sufficient for other distributions.

BIVARIATE, PART, AND PARTIAL CORRELATIONS

The different types of correlations indicated in Table 3.1 are considered bivariate correlations, or associations between two variables. Cohen and Cohen (1983), in describing correlation research, further presented the correlation between two variables controlling for the influence of a third variable. These

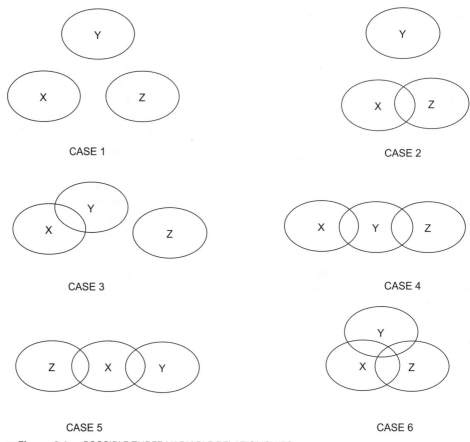

▨ **Figure 3.1:** POSSIBLE THREE-VARIABLE RELATIONSHIPS

correlations are referred to as *part* and *partial* correlations, depending upon how variables are controlled or partialled out. Some of the various ways in which three variables can be depicted are illustrated in Figure 3.1. The diagrams illustrate different situations among variables where (a) all the variables are uncorrelated (Case 1), (b) only one pair of variables is correlated (Cases 2 and 3), (c) two pairs of variables are correlated (Cases 4 and 5), and (d) all of the variables are correlated (Case 6). It is obvious that with more than three variables the possibilities become overwhelming. It is therefore important to have a theoretical perspective to suggest how certain variables are correlated in a study. A theoretical perspective is essential in specifying a model and forms the basis for testing a model in SEM.

The *partial correlation coefficient* measures the association between two variables while controlling for a third variable, for example, the association between

Table 3.5: Correlation Matrix ($n = 100$)

Variable	Age	Comprehension	Reading Level
1. Age	1.00		
2. Comprehension	.45	1.00	
3. Reading level	.25	.80	1.00

age and comprehension, controlling for reading level. Controlling for reading level in the correlation between age and comprehension partials out the correlation of reading level with age and the correlation of reading level with comprehension. *Part correlation*, in contrast, is the correlation between age and comprehension with reading level controlled for, where only the correlation between comprehension and reading level is removed before age is correlated with comprehension.

Whether a part or partial correlation is used depends on the specific model or research question. Convenient notation helps distinguish these two types of correlation (1 = age, 2 = comprehension, 3 = reading level): partial correlation, $r_{12.3}$, part correlation, $r_{1(2.3)}$ or $r_{2(1.3)}$. Different correlation values are computed depending on which variables are controlled or partialled out. For example, using the correlations in Table 3.5, we can compute the partial correlation coefficient $r_{12.3}$ (correlation between age and comprehension, controlling for reading level) as

$$r_{12.3} = \frac{r_{12} - r_{13}r_{23}}{\sqrt{(1 - r_{13}^2)(1 - r_{23}^2)}}$$

$$= \frac{.45 - (.25)\,(.80)}{\sqrt{[1 - (.25)^2]\,[1 - (.80)^2]}} = .43$$

SUPPRESSOR VARIABLE

The partial correlation coefficient should be smaller in magnitude than the Pearson product-moment correlation between age and comprehension, which is $r_{12} = .45$. If the partial correlation coefficient is not smaller than the Pearson product-moment correlation, then a *suppressor variable* may be present (Pedhazur, 1997). A suppressor variable correlates near zero with a dependent variable but correlates significantly with other predictor variables. This correlation situation serves to control for variance shared with predictor variables and not the dependent variable. The partial correlation coefficient increases in magnitude once this effect is removed from the correlation between two predictor variables with a criterion. Partial correlations will be greater in magnitude than part correlations, except when independent variables are zero correlated with the dependent variable; then, part correlations are equal to partial correlations.

The part correlation coefficient $r_{1(2.3)}$, or correlation between age and comprehension where reading level is controlled for in comprehension only, is computed as

$$r_{1(2.3)} = \frac{r_{12} - r_{13}r_{23}}{\sqrt{1 - r_{23}^2}} = \frac{.45 - (.25)(.80)}{\sqrt{1 - .80^2}} = .42$$

or in the case of correlating comprehension with age where reading level is controlled for age only is

$$r_{2(1.3)} = \frac{r_{12} - r_{13}r_{23}}{\sqrt{1 - r_{13}^2}} = \frac{.45 - (.25)(.80)}{\sqrt{1 - .25^2}} = .26.$$

The correlation, whether zero-order (bivariate), part, or partial can be tested for significance, interpreted as variance accounted for by squaring each coefficient, and diagrammed using Venn or Ballentine figures to conceptualize their relationships. In our example, the zero-order relationships among the three variables can be diagrammed as in Figure 3.2. However, the partial correlation of age with

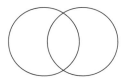

Age and Comprehension

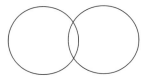

Age and Reading

Reading and Comprehension

■ **Figure 3.2:**   BIVARIATE CORRELATION

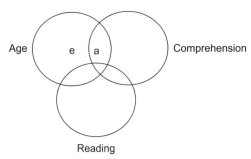

Figure 3.3: PARTIAL CORRELATION AREA

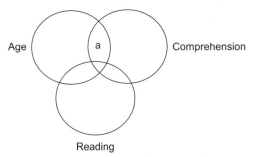

Figure 3.4: PART CORRELATION AREA

comprehension controlling for reading level would be $r_{12.3} = .43$, or area a divided by the combined area of a and e [$a/(a + e)$]; see Figure 3.3. A part correlation of age with comprehension while controlling for the correlation between reading level and comprehension would be $r_{1(2.3)} = .42$, or just area a; see Figure 3.4.

These examples consider only controlling for one variable when correlating two other variables (partial), or controlling for the impact of one variable on another before correlating with a third variable (part). Other higher-order part correlations and partial correlations are possible (for example, $r_{12.34}$, $r_{12(3.4)}$), but are beyond the scope of this book. Readers should refer to the following references for a more detailed discussion of part and partial correlation (Cohen & Cohen, 1983; Hinkle, Wiersma & Jurs, 2003; Lomax & Hahs-Vaughn, 2012; Pedhazur, 1997).

CORRELATION VERSUS COVARIANCE

The type of data matrix used for computations in structural equation modeling programs is a variance–covariance matrix. A variance–covariance matrix is made up of variance terms on the diagonal and covariance terms on the off-diagonal. If a correlation matrix is used as the input data matrix, most of the computer programs by default convert it to a variance–covariance matrix using the means and standard deviations of the variables, unless specified otherwise. The researcher has the option to input raw data, a correlation matrix, or a variance–covariance matrix. If a correlation matrix is input with variable means and standard deviations, then a variance–covariance matrix is used with unstandardized output. If only a correlation matrix is input, the means and standard deviations, by default, are set at 0 and 1, respectively, and standardized output is printed. When raw data are input, a variance–covariance matrix is computed and used.

The number of distinct elements in a variance–covariance matrix S is $p(p + 1)/2$, where p is the number of observed variables. For example, the variance–covariance matrix for the following three variables, X, Y, and Z, is as:

		X	Y	Z
	X	15.80		
$S=$	Y	10.16	11.02	
	Z	12.43	9.23	15.37

It has $3(3 + 1)/2 = 6$ distinct values: 3 variance terms in the diagonal and 3 covariance terms in the off-diagonal of the matrix.

Correlation is computed using the variance and covariance among the bivariate variables, using the following formula:

$$r = \frac{s_{XY}^2}{\sqrt{s_X^2 * s_Y^2}}$$

Dividing the covariance between two variables (covariance terms are the off-diagonal values in the matrix) by the square root of the product of the two variable variances (variances of variables are on the diagonal of the matrix) yields the following correlations among the three variables:

$r_{xy} = 10.16/(15.80 * 11.02)^{1/2} = .77$
$r_{xz} = 12.43/(15.80 * 15.37)^{1/2} = .80$
$r_{yz} = 9.23/(11.02 * 15.37)^{1/2} = .71.$

Structural equation software uses the variance–covariance matrix rather than the correlation matrix because Boomsma (1983) found that the analysis of correlation matrices led to imprecise parameter estimates and standard errors of the parameter estimates in a structural equation model. In SEM, incorrect estimation of the standard errors for the parameter estimates could lead to statistically significant parameter estimates and an incorrect interpretation of the model, that is, the parameter divided by the standard error indicates a test statistic (t or z value), which can be compared to tabled critical t values for statistical significance at different alpha levels (Table A.2). Browne (1982), Jennrich and Thayer (1973), and Lawley and Maxwell (1971) have suggested corrections for the standard errors when correlations or standardized coefficients are used in SEM. In general, a variance–covariance matrix should be used in structural equation modeling, although some SEM models require variable means, for example, structured means models which test differences in latent variable means.

VARIABLE METRICS (STANDARDIZED VS UNSTANDARDIZED)

Researchers have debated the use of unstandardized or standardized variables. The standardized coefficients are thought to be sample-specific and not stable across different samples because of changes in the variance of the variables. The unstandardized coefficients permit an examination of change across different samples. The standardized coefficients are useful, however, in determining the relative importance of each variable to other variables for a given sample. Other reasons for using standardized variables are that variables are on the same scale of measurement, are more easily interpreted, and can easily be converted back to the raw scale metric. SEM software provides the option to produce both standardized and unstandardized output. It is recommended that you report both types of output, especially because standard errors are used to compute the statistical significance of the unstandardized parameter estimates.

CAUSATION ASSUMPTIONS AND LIMITATIONS

The Pearson correlation coefficient is a measure of linear relation between two variables. The Pearson correlation coefficient establishes an inference between the values of the two variables, not a cause-and-effect relation. An experimental design that randomly samples and randomly assigns subjects to a treatment group and a control group forms the basis for making a cause-and-effect outcome decision. More specifically, the treatment effect causes a change in the outcome.

In path analysis, certain conditions are necessary for cause-and-effect outcomes (Tracz, 1992): (a) temporal order (X precedes Y in time), (b) existence of covariance or correlation between X and Y, and (c) control for other causes, for example, partial Z out of X and Y. These three conditions may not be present in the research design setting, and in such a case, only association rather than causation can be inferred. However, if *manipulative* variables are used in the study, then a researcher could change or manipulate one variable in the study and examine subsequent effects on other variables, thereby determining cause-and-effect relationships (Halpern & Pearl, 2005a; 2005b; Resta & Baker, 1972). In structural equation modeling, the amount of influence rather than a cause-and-effect relationship is assumed and interpreted by direct, indirect, and total effects among variables. Path models were initially called causal models, which created much confusion regarding inference versus causation.

Philosophical differences exist between assuming causal versus inference relationships among variables, and the resolution of these issues requires a sound theoretical perspective. Bullock, Harlow, and Mulaik (1994) provided an in-depth discussion of causation issues related to structural equation modeling research. We feel that structural equation models will evolve beyond model fit into the domain of model testing as witnessed by the many new SEM model applications today. Model testing rather than model fit can involve testing the significance of parameters, parameter change, or other factors that affect the model outcome values, and whose effects can be assessed. This approach, we believe, best depicts a causal assumption. In addition, structural models in longitudinal research can depict changes in latent variables over time (Collins & Horn, 1992). Pearl (2009) has renewed the discussion about causality, and firmly believes it is not mystical or metaphysical, but rather can be understood in terms of processes (models) that can be expressed in mathematical expressions ready for computer analysis.

SUMMARY

We have described some of the basic correlation concepts underlying structural equation modeling. This discussion included various types of bivariate correlation coefficients, part and partial correlation, variable metrics, factors affecting correlation, the assumptions required in SEM, and the causation versus inference debate in SEM modeling.

Commercial software may not compute all the types of correlation coefficients used in statistics, so the reader should refer to a standard statistics textbook for computational formulas and understanding (Hinkle, Wiersma, & Jurs, 2003). Structural equation modeling programs use a variance–covariance matrix, and include features to output the type of matrices they use. In SEM, categorical and/or ordinal variables with underlying continuous latent-variable attributes have been used with tetrachoric or polychoric correlations (Muthén, 1982, 1983, 1984; Muthén & Kaplan, 1985). SEM software permits the use of different types of correlation matrices, the creation of an asymptotic covariance matrix for input into structural equation modeling programs, and FIML estimation in the presence of missing data (Jöreskog & Sörbom, 2012). The use of various correlation coefficients and subsequent conversion into a variance–covariance matrix will continue to play a major role in structural equation modeling applications.

The chapter also presented numerous factors that affect the Pearson correlation coefficient, for example, restriction of range in the scores, outliers, skewness, and non-normality. SEM software also converts correlation matrices with standard deviations into a variance–covariance matrix, but if attenuated correlations are greater than 1.0, a non-positive definite error message will occur because of an inadmissible solution. Non-positive definite error messages are all too common among beginners because they do not screen the data, thinking instead that structural equation modeling will be unaffected. Another major concern is when OLS initial estimates lead to bad start values for the coefficients in a model; however, changing the number of default iterations sometimes solves this problem. We offer suggestions in Table 3.6 to guide how to handle these types of issues that affect correlation in SEM.

Table 3.6: Factors Affecting Correlation with Suggestions

Issue	Suggestions
Measurement scale	Need to take the measurement scale of the variables into account when computing correlations.
Restriction of range	Need to consider range of values obtained for variables, as restricted range of one or more variables can reduce the magnitude of correlations. Can consider data transformations for non-normal data.
Missing data	Need to consider missing data on one or more subjects for one or more variables as these can affect SEM results. Cases are lost with listwise deletion, pairwise deletion is often problematic (for example, different sample sizes), and thus modern methods are recommended.
Outliers	Need to consider outliers as they can affect correlations. They can either be explained, deleted, or accommodated (using either robust statistics or obtaining additional data to fill-in). Can be detected by methods such as box plots, scatterplots, histograms, or frequency distributions.
Linearity	Need to consider whether variables are linearly related, as non-linearity can reduce the magnitude of correlations. Can be detected by scatterplots and dealt with by transformations or deleting outliers.
Correction for attenuation	Less than perfect reliability on observed measures can reduce the magnitude of correlations and lead to a non-positive definite error message. Best to use multiple, high-quality measures.
Non-positive definite matrices	Can occur in a correlation or covariance matrix due to a variable that is a linear combination of other variables, collinearity, sample size less than the number of variables, negative or zero variances, correlations outside of the permissible range, or bad start values. Solutions include eliminating the bad variables, rescaling variables, and using more reasonable starting values.
Sample size	Small samples can reduce power and precision of parameter estimates. At least 100 to 150 cases are necessary for smaller models with well-behaved data.

EXERCISES

1. Given the Pearson correlation coefficients $r_{12} = .6$, $r_{13} = .7$, and $r_{23} = 4$, compute the part and partial correlations $r_{12.3}$ and $r_{1(2.3)}$.
2. Compare the variance explained in the bivariate, partial, and part correlations of Exercise 1.
3. Explain causation and describe when a cause-and-effect relationship might exist.
4. Given the following variance–covariance matrix, compute the Pearson correlation coefficients: r_{XY}, r_{XZ}, and r_{YZ}.

	X	Y	Z
X	15.80		
Y	10.16	11.02	
Z	12.43	9.23	15.37

5. Enter the correlation matrix in a statistics package ($N = 209$). Calculate the determinant of the matrix, eigenvalues, and eigenvectors. For example, R uses det() and eigen() functions. Report and interpret these values.

Correlation Matrix							
Academic	1.00						
Athletic	.43	1.00					
Attract	.50	.48	1.00				
GPA	.49	.22	.32	1.00			
Height	.10	−.04	−.03	.18	1.00		
Weight	.04	.02	−.16	−.10	.34	1.00	
Rating	.09	.14	.43	.15	−.16	−.27	1.00
S.D.	.16	.07	.49	3.49	2.91	19.32	1.01
Means	.12	.05	.42	10.34	.00	94.13	2.65

REFERENCES

Anderson, J. C., & Gerbing, D. W. (1988). Structural equation modeling in practice: A review and recommended two step approach. *Psychological Bulletin*, 103, 411–423.

Anderson, C., & Schumacker, R. E. (2003). A comparison of five robust regression methods with ordinary least squares regression: Relative efficiency, bias, and test of the null hypothesis. *Understanding Statistics*, 2, 77–101.

Arbuckle, J. L. (1996). Full information estimation in the presence of incomplete data. In G. A. Marcoulides, & R. E. Schumacker (Eds.), *Advanced structural equation modeling* (pp. 243–277). Mahwah, NJ: Lawrence Erlbaum Associates.

Beale, E. M. L., & Little, R. J. (1975). Missing values in multivariate analysis. *Journal of the Royal Statistical Society Series B,* 37, 129–145.

Bentler, P. M., & Chou, C. (1987). Practical issues in structural equation modeling. *Sociological Methods and Research,* 16, 78–117.

Boomsma, A. (1982). The robustness of LISREL against small sample sizes in factor analysis models. In K. G. Jöreskog, & H. Wold (Eds.), *Systems under indirect observation: Causality, structure, prediction (Part I)* (pp. 149–173). Amsterdam: North-Holland.

Boomsma, A. (1983). *On the robustness of LISREL against small sample size and nonnormality.* Amsterdam: Sociometric Research Foundation.

Browne, M. W. (1982). *Covariance structures.* In D. M. Hawkins (Ed.), *Topics in applied multivariate analysis* (pp. 72–141). Cambridge: Cambridge University Press.

Bullock, H. E., Harlow, L. L., & Mulaik, S. A. (1994). Causation issues in structural equation modeling. *Structural Equation Modeling: A Multidisciplinary Journal,* 1, 253–267.

Cohen, J., & Cohen, P. (1983). *Applied multiple regression/correlation analysis for the behavioral sciences* (2nd edn.). Hillsdale, NJ: Lawrence Erlbaum Associates.

Collins, L. M., & Horn, J. L. (Eds.). (1992). *Best methods for the analysis of change: Recent advances, unanswered questions, future directions.* Washington, DC: American Psychological Association.

Costello, A. B., & Osborne, J. (2005). Best practices in exploratory factor analysis: Four recommendations for getting the most from your analysis. *Practical Assessment Research & Evaluation,* 10(7), 1–9.

Crocker, L., & Algina, J. (1986). *Introduction to classical and modern test theory.* New York: Holt, Rinehart & Winston.

Davey, A., & Savla, J. (2009). *Statistical power analysis with missing data: A structural equation modeling approach.* New York: Routledge, Taylor & Francis Group.

Ding, L., Velicer, W. F., & Harlow, L. L. (1995). Effects of estimation methods, number of indicators per factor, and improper solutions on structural equation modeling fit indices. *Structural Equation Modeling: A Multidisciplinary Journal,* 2, 119–143.

Enders, C. K. (2013). Analyzing structural equation models with missing data. In G. R. Hancock, & R. O. Mueller (Eds.), *Structural equation modeling: A second course* (2nd edn.). Greenwich, CT: Information Age.

Ferguson, G. A., & Takane, Y. (1989). *Statistical analysis in psychology and education* (6th edn, pp. 493–519). New York: McGraw-Hill.

Halpern, J. Y., & Pearl, J. (2005a). Causes and explanations: A structural-model approach – Part I: Causes. *British Journal of Philosophy of Science,* 56, 843–887.

Halpern, J. Y., & Pearl, J. (2005b). Causes and explanations: A structural-model approach – Part II: Explanations. *British Journal of Philosophy of Science,* 56, 889–911.

Hinkle, D. E., Wiersma, W., & Jurs, S. G. (2003). *Applied statistics for the behavioral sciences* (5th edn.). Boston: Houghton Mifflin.

Ho, K., & Naugher, J. R. (2000). Outliers lie: An illustrative example of identifying outliers and applying robust methods. *Multiple Linear Regression Viewpoints*, 26(2), 2–6.

Hoelter, J. W. (1983). The analysis of covariance structures: Goodness-of-fit indices. *Sociological Methods and Research*, 11, 325–344.

Hu, L., Bentler, P. M., & Kano, Y. (1992). Can test statistics in covariance structure analysis be trusted? *Psychological Bulletin*, 112, 351–362.

Huber, P. J. (1981). *Robust statistics.* New York: Wiley.

Jennrich, R. I., & Thayer, D. T. (1973). A note on Lawley's formula for standard errors in maximum likelihood factor analysis. *Psychometrika*, 38, 571–580.

Jöreskog, K. G., & Sörbom, D. (2012). Some New Features in LISREL9. Chicago: Scientific Software International.

Lawley, D. N., & Maxwell, A. E. (1971). *Factor analysis as a statistical method.* London: Butterworth.

Little, R. J., & Rubin, D. B. (1987). *Statistical analysis with missing data.* New York: Wiley.

Little, R. J., & Rubin, D. B. (1990). The analysis of social science data with missing values. *Sociological Methods and Research*, 18, 292–326.

Lomax, R. G., & Hahs-Vaughn, D. L. (2012). *An introduction to statistical concepts* (3rd edn.). New York: Routledge Press (Taylor & Francis Group).

McCall, C. H., Jr. (1982). *Sampling statistics handbook for research.* Ames: Iowa State University Press.

McKnight, P. E., McKnight, K. M., Sidani, S., & Aurelio, J. F. (2007). *Missing data: A gentle introduction.* New York: Guilford.

Muthén, B. (1982). A structural probit model with latent variables. *Journal of the American Statistical Association*, 74, 807–811.

Muthén, B. (1983). Latent variable structural equation modeling with categorical data. *Journal of Econometrics*, 22, 43–65.

Muthén, B. (1984). A general structural equation model with dichotomous, ordered categorical, and continuous latent variable indicators. *Psychometrika*, 49, 115–132.

Muthén, B., & Kaplan, D. (1985). A comparison of some methodologies for the factor analysis of non-normal Likert variables. *British Journal of Mathematical and Statistical Psychology*, 38, 171–189.

Pearl, J. (2009). *Causality: Models, reasoning, and inference* (2nd edn.). Cambridge University Press: London.

Pearson, K. (1896). Mathematical contributions to the theory of evolution. Part 3. Regression, heredity and panmixia. *Philosophical Transactions, A*, 187, 253–318.

Pedhazur, E. J. (1997). *Multiple regression in behavioral research: Explanation and prediction* (3rd edn.). Fort Worth: Harcourt Brace.

Peng, C.-Y. J., Harwell, M., Liou, S.-M., & Ehman, L. H. (2007). Advances in missing data methods and implications for educational research. In S. S. Sawilowsky (Ed.), *Real data analysis.* Charlotte: Information Age.

Resta, P. E., & Baker, R. L. (1972). *Selecting variables for educational research.* Inglewood, CA: Southwest Regional Laboratory for Educational Research and Development.

Rousseeuw, P. J., & Leroy, A. M. (1987). *Robust regression and outlier detection*. New York: Wiley.

Schafer, J. L. (1997). *Analysis of incomplete data*. Boca Raton, FL: Chapman & Hall/CRC Press.

Spearman, C. (1904). The proof and measurement of association between two things. *American Journal of Psychology, 15*, 72–101.

Staudte, R. G., & Sheather, S. J. (1990). *Robust estimation and testing*. New York: Wiley.

Stevens, S. S. (1968). Measurement, statistics, and the schempiric view. *Science, 161*, 849–856.

Tankard, J. W., Jr. (1984). *The statistical pioneers*. Cambridge, MA: Schenkman.

Tracz, S. M. (1992). The interpretation of beta weights in path analysis. *Multiple Linear Regression Viewpoints, 19*(1), 7–15.

Wothke, W. (1993). Nonpositive definite matrices in structural equation modeling. In K. A. Bollen, & S. J. Long (Eds.), *Testing structural equation models* (pp. 256–293). Newbury Park, CA: Sage.

Wothke, W. (2000). Longitudinal and multi-group modeling with missing data. In T. D. Little, K. U. Schnabel, & J. Baumert (Eds.), *Modeling longitudinal and multiple group data: Practical issues, applied approaches and specific examples* (pp. 1–24). Mahwah, NJ: Lawrence Erlbaum Associates.

Chapter 4

REGRESSION MODELS

CHAPTER CONCEPTS

Explanation versus prediction
Standardized partial regression coefficients
Coefficient of determination
Squared multiple correlation coefficient
Full versus restricted models
Measurement error
Additive versus relational model

Multiple regression models predict observed dependent variable values from knowledge of multiple independent observed variable values. Multiple regression, a general linear modeling approach to the analysis of data, has become increasingly popular since 1967 (Bashaw & Findley, 1968). In fact, it has become recognized as an approach that bridges the gap between correlation and analysis of variance in answering research hypotheses (McNeil, Kelly, & McNeil, 1975). Many statistical textbooks have elaborated the relationship between multiple regression and analysis of variance (Draper & Smith, 1966; Edwards, 1979; Hinkle, Wiersma, & Jurs, 2003). Graduate students who take an advanced statistics course are typically provided with the multiple linear regression framework for data analysis. Given knowledge of multiple regression techniques, an understanding can be extended to various multivariable statistical techniques (Newman, 1988).

This chapter shows how beta weights (standardized partial regression coefficients) are computed in multiple regression equations using structural equation modeling software. More specifically, we illustrate how the structural equation modeling approach can be used to compute parameter estimates in multiple regression and what types of output are reported. We begin with a brief overview of multiple regression concepts.

OVERVIEW

Multiple regression techniques require a basic understanding of sample statistics (sample size, mean, and variance), standardized variables, correlation, and partial correlation (Cohen & Cohen, 1983; Houston & Bolding, 1974; Pedhazur, 1982). In standard-score form (z scores), the simple linear regression equation for predicting the dependent variable Y from a single independent variable X is

$$\hat{z}_y = \beta\, z_x,$$

where β is the standardized regression coefficient. The basic rationale for using the standard-score formula is that variables are converted to the same scale of measurement, the z scale. Conversion back to the raw-score scale is easily accomplished by using the sample mean and the standard deviation.

The relationship connecting the Pearson product-moment correlation coefficient, the unstandardized regression coefficient b and the standardized regression coefficient β is

$$\beta = \frac{\sum z_x z_y}{\sum z_x^2} = b\,\frac{s_x}{s_y} = r_{xy},$$

where s_x and s_y are the sample standard deviations for variables X and Y, respectively. For two independent variables, the multiple linear regression equation with standard scores is

$$\hat{z}_y = \beta_1 z_1 + \beta_2 z_2$$

and the standardized partial regression coefficients β_1 and β_2 are computed from

$$\beta_1 = \frac{r_{y1} - r_{y2} r_{12}}{1 - r_{12}^2} \quad \text{and} \quad \beta_2 = \frac{r_{y2} - r_{y1} r_{12}}{1 - r_{12}^2}.$$

The correlation between the dependent observed variable Y and the predicted scores $\hat{Y}$ is given the special name *multiple correlation coefficient*. It is written as

$$R_{y\hat{y}} = R_{y.12},$$

where the latter subscripts indicate that the dependent variable Y is being predicted by two independent variables, X_1 and X_2. The *squared multiple correlation coefficient* is computed as

$$R_{Y\hat{Y}}^2 = R_{y.12}^2 = \beta_1 r_{y1} + \beta_2 r_{y2}.$$

The squared multiple correlation coefficient indicates the amount of variance explained, predicted, or accounted for in the dependent variable by the set of independent predictor variables. The R^2 value is also interpreted as an effect size or model-fit criterion in multiple regression analysis.

Kerlinger and Pedhazur (1973) indicated that multiple regression analysis can play an important role in prediction and explanation. Prediction and explanation reflect different research questions, study designs, inferential approaches, analysis strategies, and reported information. In prediction, the main emphasis is on practical application such that independent variables are chosen by their effectiveness in enhancing prediction of the dependent variable values. In explanation, the main emphasis is on the variability in the dependent variable explained by a theoretically meaningful set of independent variables. Huberty (2003) established a clear distinction between prediction and explanation when referring to multiple correlation analysis (MCA) and multiple regression analysis (MRA). In MCA, a parameter of interest is the correlation between the dependent variable Y and a composite of the independent variables X_p. The adjusted formula using sample size n and the number of independent predictors, p, is

$$R^2_{Adj} = R^2 - \frac{p}{n-p-1}(1-R^2).$$

In MRA, regression weights are also estimated to achieve a composite for the independent variables X_p, but the index of fit R^2 is computed differently as

$$R^2_{Adj*} = R^2 - \frac{2p}{n-p}(1-R^2).$$

When comparing these two formulas, we see that R^2_{Adj*} has a larger adjustment. For example, given $R^2 = .50$, $p = 10$ predictor variables and $n = 100$ subjects, these two different fit indices are

$$R^2_{Adj} = R^2 - \frac{p}{n-p-1}(1-R^2) = .50 - .11(.50) = .50 - .055 = .45$$

$$R^2_{Adj*} = R^2 - \frac{2p}{n-p}(1-R^2) = .50 - .22(.50) = .50 - .11 = .39.$$

Hypothesis testing would involve using the *expected value* or chance value of R^2 for testing the null hypothesis, which is $p/(n-1)$, not 0 as typically indicated. In our example, the expected or chance value for $R^2 = 10/99 = .10$, so the null hypothesis is H_0: $\rho^2 = .10$. An F test used to test the statistical significance of the R^2 value is

$$F = \frac{R^2/p}{(1-R^2)/n-p-1}.$$

In our example,

$$F = \frac{R^2/p}{(1-R^2)/n-p-1} = \frac{.50/10}{(1-.50)/89} = \frac{.05}{.0056} = 8.9,$$

which is statistically significant when compared to the tabled $F = 1.93$, $df = 10$; 89, $p < .05$ (Table A.5). In addition to the statistical significance test, a researcher should calculate *effect sizes* and *confidence intervals* to aid understanding and interpretation (Soper, 2010).

The *effect size* (ES) is computed as ES $= R^2 - [p/(n - 1)]$. In our example, ES $R^2_{Adj} = .45 - .10 = .35$ and ES $R^2_{Adj*} = .39 - .10 = .29$. This indicates a moderate to large effect size according to Cohen (1988), who gave a general reference for effect sizes (small $= .1$, medium $= .25$, and large $= .4$).

Confidence intervals (CIs) around the R^2 value can also help our interpretation of multiple regression analysis. Steiger and Fouladi (1992) reported an R^2 CI DOS program that computes confidence intervals, power, and sample size. Steiger and Fouladi (1997) and Cumming and Finch (2001) both discussed the importance of converting the central F value to an estimate of the non-central F before computing a confidence interval around R^2. Smithson (2001) wrote an R^2 SPSS program to compute confidence intervals. Confidence intervals around R^2 values, however, have not been adopted by researchers and thus are not reported in published research.

After assessing our initial regression model fit, we might want to determine whether adding or deleting an independent variable would improve the index of fit R^2, but we avoid using stepwise regression methods (Huberty, 1989). We run a second multiple regression equation where a single independent variable is added or deleted to obtain a second R^2 value. We then compute a different F test to determine the statistical significance between the two regression models as follows

$$F = \frac{(R^2_F - R^2_R)/(p_1 - p_2)}{(1-R^2_F)/n - p_1 - 1},$$

where R^2_F is from the multiple regression equation with the full original set of independent variables p_1 and R^2_R is from the multiple regression equation with the reduced set of independent variables p_2. In our heuristic example, we drop a single independent variable and obtain $R^2_R = .49$ with $p_2 = 9$ predictor variables. The F test is computed as

$$F = \frac{(R_F^2 - R_R^2)/(p_1 - p_2)}{(1 - R_F^2)/n - p_1 - 1} = \frac{(.50 - .49)/(10 - 9)}{(1 - .50)/100 - 10 - 1} = \frac{.01}{.0056} = 1.78.$$

The F value is not significant at the .05 level, so the variable we dropped does not statistically add to the prediction of Y, which supports our dropping the single predictor variable; that is, a 1% decrease in R^2 is not statistically significant. The nine-variable regression model therefore provides a more parsimonious model.

It is important to understand the basic concepts of multiple regression and correlation because they provide a better understanding of hypothesis testing, prediction, and explanation of a dependent variable. A review of multiple regression techniques also helps us to better understand path analysis, and structural equation modeling in general. An SEM example is presented next to further clarify these basic multiple regression computations.

MULTIPLE REGRESSION EXAMPLE

The multiple linear regression analysis is conducted using data from Chatterjee and Yilmaz (1992). The data file contains scores from 24 patients on four variables (Var1 = patient's age in years, Var2 = severity of illness, Var3 = level of anxiety, and Var4 = satisfaction level). Given raw data, two different approaches are possible: (a) read in raw data file, or (b) compute a correlation or variance–covariance matrix for input into the software. We choose to compute and input a variance–covariance matrix into the software program. SEM software does not output all of the same information and related diagnostic results for multiple regression that you may be accustomed to viewing in SAS, SPSS, STATA, etc.

A regression equation is a theoretical model specified by the researcher. Therefore, the *model specification* involves finding relevant theory and prior research to formulate a theoretical regression model. The researcher is interested in specifying a regression model that should be confirmed with sample variance–covariance data, thus yielding a high R^2 value and statistically significant F value. Model specification directly involves deciding which variables to include or not to include in the theoretical regression model.

If the researcher does not select the right variables, then the regression model could be misspecified and lack validity (Tracz, Brown, & Kopriva, 1991). The problem is that a misspecified model may result in biased parameter estimates or estimates that are systematically different from what they are in the true population model. This bias is known as *specification error*.

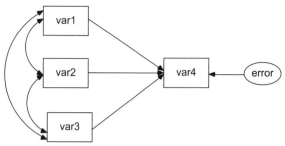

■ **Figure 4.1:** SATISFACTION REGRESSION MODEL

The researcher's goal is to determine whether the theoretical regression model fits the sample variance–covariance structure in the data, that is, whether the sample variance–covariance matrix implies some underlying theoretical regression model. The multiple regression model of theoretical interest in our example is to predict the satisfaction level of patients based on patient's age, severity of illness, and level of anxiety (independent variables). This would be characteristic of an MCA model because a particular set of variables was selected based on theory. The dependent variable var4 is therefore predicted by the three independent variables (var1, var2, and var3). The diagram of the implied regression model is shown in Figure 4.1. The curved arrows indicate correlations between the observed independent variables. The lines pointing toward var4 indicate the direct paths for regression weights. The oval with error indicates the $1 - R^2$ or unexplained variance not accounted for by the three independent predictor variables.

Once a theoretical regression model is specified, the next concern is *model identification*. Model identification refers to deciding whether a set of unique parameter estimates can be computed for the regression equation. Algebraically, every free parameter in the multiple regression equation can be estimated from the sample variance–covariance matrix (a free parameter is an unknown parameter that you want to estimate). The number of distinct values in the sample variance–covariance matrix equals the number of parameters to be estimated; thus, multiple regression models are always considered *just-identified models* because all parameters are estimated.

SEM computer output will always indicate that regression analyses are *saturated* models; that is, $\chi^2 = 0$ and degrees of freedom = 0. The regression model includes three independent variable variances, three covariance terms, three regression weights for the independent variables, and one error term, so all parameters in the regression equation are being estimated. Traditional software (SAS, SPSS, etc.) reports the R^2 and F values, whereas SEM software reports a chi-square value, because SEM software is testing the difference between the original sample variance–covariance matrix and the model-implied variance–covariance matrix given the regression equation.

The estimation of the regression weights is called *model estimation*, that is, computing the sample regression weights for the independent predictor variables. The term *model estimation* is used because there are several different estimation methods. The most common estimation method is ordinary least squares estimation (*unweighted least squares*; ULS), which selects the regression weights based on minimizing the sum of squared errors. However, there are other estimation methods used in statistics, especially SEM software, for example: *maximum likelihood* (ML) estimation, *two-stage least squares* (2SLS), *weighted least squares* (WLS), *generalized least squares* (GLS), *diagonally weighted least squares* (DWLS), and robust versions in Mplus (MLM, MLMV, WLSM, WLSMV). The different estimation methods were derived to be used under various data analysis situations, including non-normality, small sample sizes, outliers, etc., to afford a more robust estimation of parameter estimates. We will be discussing these other estimation methods when we present the different SEM modeling approaches later in the book.

The squared multiple correlation with three predictor variables predicting the dependent variable Y is

$$R^2_{y.123} = \beta_1 r_{y1} + \beta_2 r_{y2} + \beta_3 r_{y3}.$$

The correlation coefficients are multiplied by their respective standardized partial regression weights and summed to yield the squared multiple regression coefficient $R^2_{y.123}$.

In LISREL9 we can write a SIMPLIS program to compute the regression weights in the regression model. The SIMPLIS program includes a *title* command, an *observed variable* command to specify variable names, *sample size* command, and *covariance matrix* command. The *equation* command specifies the regression equation with the dependent variable on the left-hand side of the equation. The *number of decimals* and *path diagram* commands are optional. The *end of problem* command ends the program.

The SIMPLIS program commands can be saved in a file (*regression.spl*) and run in the free student version of the software (www.ssicentral.com). The basic program setup is:

```
Regression Analysis Example (no intercept term)
Observed variables: VAR1 VAR2 VAR3 VAR4
Sample size: 24
Covariance matrix:
91.384
30.641 27.288
0.5840.641 0.100
-122.616 -52.576 -2.399 281.210
Equation: VAR4 = VAR1 VAR2 VAR3
Number of decimals = 3
Path Diagram
End of Problem
```

The regression output without an *intercept* term in the regression equation is:

```
VAR4 = - 1.153*VAR1 - 0.267*VAR2 - 15.546*VAR3, Errorvar.= 88.515
Standerr (0.273)    (0.533)      (7.080)         (27.402)
Z-values -4.218     -0.501       -2.196          3.230
P-values  0.000      0.616        0.028          0.001

Goodness-of-Fit Statistics
Degrees of Freedom for (C1) - (C2) = 0
Maximum Likelihood Ratio Chi-Square (C1) = 0.0 (P = 1.000)
Browne's (1984) ADF Chi-Square (C2_N2) = 0.00 (P = 1.000)
The model is saturated, the fit is perfect!
```

The regression weights are listed in front of each independent variable (VAR1, VAR2, VAR3). Below each regression weight is the standard error in parenthesese; for example, VAR1 regression weight has a standard error of .273, with the Z value indicated below that, and a p value listed below the Z value. The Z value is computed as the parameter estimate divided by the standard error ($Z = -1.153/.273 = -4.128$). If testing each regression weight at the critical $z = 1.96$, $\alpha = .05$ level of significance, then VAR1 and VAR3 are statistically significant, but VAR2 is not ($Z = -.501$). The $R^2 = .685$ or 69% of the variability in Y scores (VAR4) is predicted by knowledge of VAR1, VAR2, and VAR3. This example is further explained in Jöreskog and Sörbom (1993, pp. 1–6).

Model testing involves determining the fit of the theoretical regression model. We can calculate by hand the R^2 value using the correlation matrix and standardized beta weights as shown in Table 4.1.

The adjusted R^2 value for the MCA theoretical regression model approach is

Table 4.1: Correlation Matrix ($n = 24$)

Correlation Matrix				
	VAR1	VAR2	VAR3	VAR4
VAR1	1.0000			
VAR2	0.6136	1.0000		
VAR3	0.1935	0.3888	1.0000	
VAR4	−0.7649	−0.6002	−0.4530	1.0000

$$R^2_{y.123} = \beta_1 r_{y1} + \beta_2 r_{y2} + \beta_3 r_{y3}$$

$$= -.657(-.7649) + -.083(-.6002) + -.294(-.4530)$$

$$= .685.$$

$$R^2_{Adj} = R^2 - \frac{p}{n-p-1}(1-R^2) = .685 - .15(.315) = .685 - .047 = .638.$$

The F test for the significance of the R^2 value is

$$F = \frac{R^2/p}{(1-R^2)/n-p-1} = \frac{.685/3}{(1-.685)/20} = \frac{.228}{.0157} = 14.52 .$$

The *effect size* is

$$R^2 - [p/(n-1)] = .685 - [3/23] = .685 - .130 = .554.$$

This would be considered a large effect size. Notice that the SEM software does not provide these values.

The results indicated that a patient's age, severity of illness, and level of anxiety make up a statistically significant set of predictors of a patient's satisfaction level. There is a large effect size so one might expect similar results when conducting a regression analysis on another sample of data. The negative standardized regression coefficients indicate that as patient age, severity of illness, and anxiety increase, a patient's satisfaction decreases. SEM software can output both standardized and unstandardized parameter estimates. Both should be reported, so we showed how they are computed. Also, we have not dropped the non-significant second variable, which is discussed next.

The theoretical regression model included a set of three independent explanatory variables, which resulted in a statistically significant $R^2 = .685$. This implied that 69% of the patient satisfaction level score variance was explained by knowledge of a patient's age, severity of illness, and level of anxiety. The regression analysis, however, indicated that the regression weight for VAR2 was not statistically different from zero ($z = -0.501$, $p = .10$). Thus, one might consider *model modification* where the theoretical regression model is modified to produce a better fitting model. In multiple regression, the two different regression equations would yield different R^2 values, thus an F test of the difference between the two R^2 values would be computed.

We would run the SIMPLIS program again, but this time drop VAR2 from the equation command. The output would now only show two independent predictor variables, and a different R^2 value.

```
    VAR4 = - 1.235*VAR1 - 16.780*VAR3, Errorvar.= 89.581, R² = 0.681
   Standerr  (0.216)      (6.517)          (27.063)
   Z-values  -5.727       -2.575            3.310
   P-values   0.000        0.010            0.001
```

The F test for a difference between the two models is

$$F = \frac{(R_F^2 - R_R^2)/(p_1 - p_2)}{(1 - R_F^2)/n - p_1 - 1} = \frac{(.685 - .681)/(3 - 2)}{(1 - .685)/24 - 3 - 1} = \frac{.004}{.016} = .25 .$$

The F test for the difference in the two R^2 values was non-significant, indicating that dropping VAR2 does not affect the explanation of a patient's satisfaction level ($R^2 = .685$ vs $R^2 = .681$). We therefore use the more parsimonious two-variable regression model (68% of the variance in a patient's satisfaction level is explained by knowledge of a patient's age and level of anxiety, that is, 68% of $281.210 = 191.22$). The F test is also not provided in the SEM software.

Because the R^2 value is not 1.0 (perfect explanation or prediction), additional variables could be added if additional research indicated that another variable was relevant to a patient's satisfaction level, for example, the number of psychological assessment visits. The unexplained error variance (89.581) was statistically significant, that is, $1 - R^2 = 1 - .68 = .32$ (32%), so additional significant predictor variables would be helpful in accounting for the unexplained variance. Obviously, more variables can be added in the model modification process, but a theoretical basis should be established by the researcher for the additional variables.

SUMMARY

A basic regression analysis was conducted in SEM. We discovered that the model-fit statistics and computer output information in SEM are not the same as in traditional statistics packages that run multiple regression. The parameter estimates can be computed using different estimation methods. The regression models are considered saturated just-identified models, because all parameters are estimated. We also showed that the selection of independent variables in the regression model (model specification) and the subsequent regression model modification are key issues not easily resolved without a sound theoretical justification.

The selection of a set of independent variables and the subsequent regression model modification are important issues in multiple regression. How does a researcher determine the best set of independent variables for explanation or prediction? It is highly recommended that a regression model be based on some theoretical framework that can be used to guide the decision of what variables to include. Model specification consists of determining what variables to include in the model and which variables are independent or dependent. A systematic determination of the most important set of variables can then be accomplished by setting the partial regression weight of a single variable to zero, thus testing full and restricted models for a difference in the R^2 values (F test). This approach and other alternative methods were presented by Darlington (1968).

In multiple regression, the selection of a wrong set of variables can yield erroneous and inflated R^2 values. The process of determining which set of variables yields the best prediction, given time, cost, and staffing, is often problematic because several methods and criteria are available to choose from. Recent methodological reviews have indicated that stepwise methods are not preferred, and that an *all-possible-subset* approach is recommended (Huberty, 1989; Thompson, Smith, Miller, & Thomson, 1991). In addition, the Mallows C_P statistic is advocated by some rather than R^2 for selecting the best set of predictors (Mallows, 1966; Schumacker, 1994; Zuccaro, 1992). Overall, which variables are included in a regression equation will determine the validity of the model and the theoretical rationale of the researcher. For example, if the intercept term is omitted, the predictor variables are compared on the same scale with the intercept value of 0. However, if an intercept term is included, then the intercept indicates a starting point or baseline measure (see chapter footnote).

Because multiple regression techniques have been shown to be robust to violations of assumptions (Bohrnstedt & Carter, 1971) and applicable to contrast coding, dichotomous coding, ordinal coding (Lyons, 1971), and criterion scaling (Schumacker, 1993), they have been used in a variety of research designs. In fact, multiple regression equations can be used to address several different types of research questions. The model specification issue, however, is paramount in achieving a valid multiple regression model. Replication, cross-validation, and bootstrapping have all been applied in multiple regression to determine the validity of a regression model.

There are other issues related to using the regression method, namely, variable measurement error and the additive nature of the regression analysis. These two issues are described next.

MEASUREMENT ERROR

The issue of unreliable variable measurements and their effect on multiple regression has been previously discussed (Cleary, 1969; Cochran, 1968; Fuller & Hidiroglou, 1978; Subkoviak & Levin, 1977; Sutcliffe, 1958). A recommended solution was to multiply the dependent variable reliability and/or average of the independent variable reliabilities by the R^2 value (Cochran, 1968, 1970). The basic equation using only the reliability of the dependent variable is

$$\hat{R}^2_{y.123} = R^2_{y.123} * r_{yy},$$

or, including the dependent variable reliability and the average of the independent variable reliabilities,

$$\hat{R}^2_{y.123} = R^2_{y.123} * r_{yy} * \bar{r}_{xx}.$$

This adjustment for unreliability is not always possible if the reliabilities of the dependent and independent variables are unknown. This correction to R^2 for measurement error (unreliability) has intuitive appeal given the definition of classical reliability, namely the proportion of true score variance accounted for given the observed scores. In our previous example, $R^2 = .68$. If the dependent variable reliability was .80, then only 54% of the variance (.80 × .68 = .54) in patient's satisfaction level is true variance, rather than 68%. Similarly, if the average of the two independent variable reliabilities was .90, then multiplying .68 by .80 by .90 yields only 49% variance as true variance. Obviously, unreliable variables (measurement error) can have a dramatic effect on statistics and our interpretation of the results. Werts, Rock, Linn, and Jöreskog (1976) examined correlations, variances, covariances, and regression weights with and without measurement error and developed a program to correct the regression weights for attenuation. The basic concern in SEM is that unreliable measured variables coupled with a potential misspecified model do not represent theory well.

The impact of measurement error on statistical analyses is not new, but is often forgotten by researchers. Fuller (1987) extensively covered structural equation modeling, and especially extended regression analysis to the case where the variables were measured with error. Cochran (1968) studied four different aspects of how measurement error affected statistics: (a) types of mathematical models, (b) standard techniques of analysis that take measurement error into account, (c) effect of errors of measurement in producing bias and reduced precision and what remedial procedures are available, and (d) techniques for studying error of measurement. Cochran (1970) also studied the effects of measurement error on the squared multiple correlation coefficient.

The validity and reliability issues in measurement have traditionally been handled by first examining the validity and reliability of scores on instruments used in a particular research design. Given an acceptable level of score validity and reliability, the scores are then used in a statistical analysis. The traditional statistical analysis of these scores using multiple regression, however, did not adjust for measurement error, so it is not surprising that an approach such as SEM was developed to incorporate measurement error adjustments into the statistical analyses (Loehlin, 1992).

ADDITIVE EQUATION

The multiple regression equation by definition is additive ($Y = X_1 + X_2$), and thus does not permit any other relations among the variables to be specified. This limits the potential for variables to have direct, indirect, and total effects on each other in a regression model. In fact, a researcher's interest should not be with the Pearson product-moment correlations per se, but rather with partial or part correlations that reflect the unique additive contribution of each variable, that is, standardized partial regression weights. Even with this emphasis, the basic problem is that variables are typically added in a regression model, a process that functions ideally only if all independent variables are highly correlated with the dependent variable and uncorrelated among themselves. Path models, in contrast, provide theoretically meaningful relations in a manner not restricted to direct effects in the additive regression model (Schumacker, 1991).

Multiple regression as a general data-analytic technique is widely accepted and used by educational researchers, behavioral scientists, and biostatisticians. Multiple regression methods basically determine the overall contribution of a set of observed variables to explanation or prediction, test full and restricted models for the significant contribution of a variable in a model, and delineate the best subset of multiple independent predictors. Multiple regression equations also permit the use of nominal, ordinal, effect, contrast, or polynomial coded variables (Pedhazur, 1982; Pedhazur & Schmelkin, 1992). The multiple regression approach, however, is not robust to measurement error and model misspecification

(Bohrnstedt & Carter, 1971), and gives an additive model rather than a relational model; hence, path models play an important role in defining more meaningful theoretical models to test.

CHAPTER FOOTNOTE

Regression Model with Intercept Term

In LISREL (Jöreskog & Sörbom, 1993) we see our first use of the CONST command, which uses a mean value and thus includes an intercept term in the model. In the LISREL9 Student Examples folder, *SPLEX*, the program **EX1A.SPL** computes the regression equation *without* an intercept term, while the program **EX1B.SPL** computes the regression equation *with* an intercept term. In general, if you include sample means, then an intercept term is included in the equation. These examples are further explained in *LISREL8: Structural Equation Modeling with the SIMPLIS Command Language* (Jöreskog & Sörbom, 1993, pp. 1–6). The **EX1B.SPL** program which uses an intercept term (CONST) is:

```
Regression of GNP
Observed variables: GNP LABOR CAPITAL TIME
Means: 180.435 45.565 50.087 13.739
Covariance Matrix:
 4256.530
 449.016 52.984
 1535.097 139.449 1114.447
 537.482 53.291 170.024 73.747
Sample Size: 23
Equation: GNP = CONST LABOR CAPITAL TIME
Path Diagram
End of Problem
```

EXERCISES

1. Analyze the regression model using the covariance matrix in Table 4.2 without an intercept term, and a sample size of 23. The theoretical regression model specifies that the dependent variable, gross national product (GNP), is predicted by labor, capital, and time (three independent variables).
2. Is there an alternative regression model that predicts GNP better? Report the F, effect size, and R^2 for the revised model. The regression model is shown in Figure 4.2.

Table 4.2: Covariance Matrix ($n = 23$)

Covariance Matrix				
GNP	4256.530			
Labor	449.016	52.984		
Capital	1535.097	139.449	1114.447	
Time	537.482	53.291	170.024	73.747

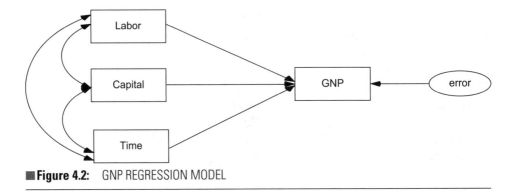

■**Figure 4.2:**   GNP REGRESSION MODEL

REFERENCES

Bashaw, W. L., & Findley, W. G. (1968). *Symposium on general linear model approach to the analysis of experimental data in educational research.* (Project No. 7-8096.) Washington, DC: U.S. Department of Health, Education, and Welfare.

Bohrnstedt, G. W., & Carter, T. M. (1971). Robustness in regression analysis. In H. L. Costner (Ed.), *Sociological methodology* (pp. 118–146). San Francisco, CA: Jossey-Bass.

Chatterjee, S., & Yilmaz, M. (1992). A review of regression diagnostics for behavioral research. *Applied Psychological Measurement*, 16, 209–227.

Cleary, T. A. (1969). Error of measurement and the power of a statistical test. *British Journal of Mathematical and Statistical Psychology,* 22, 49–55.

Cochran, W. G. (1968). Errors of measurement in statistics. *Technometrics,* 10, 637–666.

Cochran, W. G. (1970). Some effects of errors of measurement on multiple correlation. *Journal of the American Statistical Association,* 65, 22–34.

Cohen, J. (1988). *Statistical power analysis for the behavioral sciences* (2nd edn.). Hillsdale, NJ: Lawrence Erlbaum.

Cohen, J., & Cohen, P. (1983). *Applied multiple regression/correlation analysis for the behavioral sciences* (2nd edn.). Hillsdale, NJ: Lawrence Erlbaum.

Cumming, G., & Finch, S. (2001). A primer on the understanding, use and calculation of confidence intervals that are based on central and noncentral distributions. *Educational and Psychological Measurement*, 61, 532–574.

Darlington, R. B. (1968). Multiple regression in psychological research and practice. *Psychological Bulletin,* 69, 161–182.

Draper, N. R., & Smith, H. (1966). *Applied regression analysis.* New York: Wiley.

Edwards, A. L. (1979). *Multiple regression and the analysis of variance and covariance.* San Francisco, CA: Freeman.

Fuller, W. A. (1987). *Measurement error models.* New York: Wiley.

Fuller, W. A., & Hidiroglou, M. A. (1978). Regression estimates after correcting for attenuation. *Journal of the American Statistical Association,* 73, 99–104.

Hinkle, D. E., Wiersma, W., & Jurs, S. G. (2003). *Applied statistics for the behavioral sciences* (5th edn.). Boston, MA: Houghton Mifflin.

Houston, S. R., & Bolding, J. T., Jr. (1974). Part, partial, and multiple correlation in commonality analysis of multiple regression models. *Multiple Linear Regression Viewpoints,* 5, 36–40.

Huberty, C. J. (1989). Problems with stepwise methods—Better alternatives. In B. Thompson (Ed.), *Advances in social science methodology* (Vol. 1, pp. 43–70). Greenwich, CT: JAI.

Huberty, C. J. (2003). Multiple correlation versus multiple regression. *Educational and Psychological Measurement,* 63, 271–278.

Jöreskog, K. G., & Sörbom, D. (1993). *LISREL8: Structural equation modeling with the SIMPLIS command language.* Chicago, IL: Scientific Software International.

Kerlinger, F. N., & Pedhazur, E. J. (1973). *Multiple regression in behavioral research.* New York: Holt, Rinehart, & Winston.

Loehlin, J. C. (1992). *Latent variable models: An introduction to factor, path, and structural analysis* (2nd edn.). Mahwah, NJ: Lawrence Erlbaum.

Lyons, M. (1971). Techniques for using ordinal measures in regression and path analysis. In H. L. Costner (Ed.), *Sociological methodology* (pp. 147–171). San Francisco, CA: Jossey-Bass.

Mallows, C. L. (1966, March). *Choosing a subset regression.* Paper presented at the Joint Meetings of the American Statistical Association, Los Angeles.

McNeil, K. A., Kelly, F. J., & McNeil, J. T. (1975). *Testing research hypotheses using multiple linear regression.* Carbondale: Southern Illinois University Press.

Newman, I. (1988, October). *There is no such thing as multivariate analysis: All analyses are univariate.* President's address at Mid-Western Educational Research Association, Chicago.

Pedhazur, E. J. (1982). *Multiple regression in behavioral research: Explanation and prediction* (2nd edn.). New York: Holt, Rinehart, & Winston.

Pedhazur, E. J., & Schmelkin, L. P. (1992). *Measurement, design, and analysis: An integrated approach.* Hillsdale, NJ: Lawrence Erlbaum.

Schumacker, R. E. (1991). Relationship between multiple regression, path, factor, and LISREL analyses. *Multiple Linear Regression Viewpoints,* 18, 28–46.

Schumacker, R. E. (1993). Teaching ordinal and criterion scaling in multiple regression. *Multiple Linear Regression Viewpoints,* 20, 25–31.

Schumacker, R. E. (1994). A comparison of the Mallows C_p and principal component regression criteria for best model selection. *Multiple Linear Regression Viewpoints,* 21, 12–22.

Smithson, M. (2001). Correct confidence intervals for various regression effect sizes and parameters: The importance of noncentral distributions in computing intervals. *Educational and Psychological Measurement*, 61, 605–632.

Soper, D. Statistics Calculators. Retrieved January 2010 from www.danielsoper.com/statcalc/.

Steiger, J. H., & Fouladi, T. (1992). R2: A computer program for interval estimation, power calculation, and hypothesis testing for the squared multiple correlation. *Behavior Research Methods, Instruments, and Computers*, 4, 581–582.

Steiger, J. H., & Fouladi, T. (1997). Noncentrality interval estimation and the evaluation of statistical models. In L. Harlow, S. Mulaik, & J. H. Steiger (Eds.), *What if there were no significance tests?* (pp. 222–257). Mahwah, NJ: Lawrence Erlbaum.

Subkoviak, M. J., & Levin, J. R. (1977). Fallibility of measurement and the power of a statistical test. *Journal of Educational Measurement*, 14, 47–52.

Sutcliffe, J. P. (1958). Error of measurement and the sensitivity of a test of significance. *Psychometrika*, 23, 9–17.

Thompson, B., Smith, Q. W., Miller, L. M., & Thomson, W. A. (1991, January). *Stepwise methods lead to bad interpretations: Better alternatives*. Paper presented at the annual meeting of the Southwest Educational Research Association, San Antonio, TX.

Tracz, S. M., Brown, R., & Kopriva, R. (1991). Considerations, issues, and comparisons in variable selection and interpretation in multiple regression. *Multiple Linear Regression Viewpoints*, 18, 55–66.

Werts, C. E., Rock, D. A., Linn, R. L., & Jöreskog, K. G. (1976). Comparison of correlations, variances, covariances, and regression weights with or without measurement error. *Psychological Bulletin*, 83, 1007–1013.

Zuccaro, C. (1992). Mallows C_p statistic and model selection in multiple linear regression. *Journal of the Market Research Society*, 34, 163–172.

Chapter 5

PATH MODELS

CHAPTER CONCEPTS

Path model diagrams
Direct and indirect effects, correlated independent variables
Path coefficients (standardized partial coefficients)
Decomposition of correlations
Original and model-implied (reproduced) matrices
Residual and standardized residual matrix

Path models extend multiple regression models by including several regression equations. Path models specify direct, indirect, and correlated effects among the observed variables in a theoretical model. Thus, path models simultaneously solve the several multiple regression equations specified in a path model. Path analysis also permits a comparison of the original sample correlation matrix to the implied matrix that is produced based on the specified path model.

Sewall Wright is credited with the development of path analysis as a method for studying the direct and indirect effects of variables (Wright, 1921, 1934, 1960). Path analysis was originally presented as a method for testing causal outcomes; however, today, it is understood as a method to test theoretical models that depict relations amongst variables, not as a *causal modeling* technique. A causal model would have to meet several criteria, which is generally not available using bivariate correlations amongst variables. A causal model would need to include the following conditions:

1. Temporal ordering of variables exists.
2. Covariation or correlation is present among variables.
3. Other causes are controlled for.
4. X is manipulated, which causes a change in Y.

Obviously, a model that is tested over time (longitudinal research) and manipulates certain variables to assess the change in other variables (experimental research) more closely approaches the idea of causation. In the social and behavioral sciences, the issue of causation is not as straightforward as in the hard sciences, but it has the potential to be modeled. Pearl (2009) argues for causation in the behavioral sciences; his rationale is that causation is a process that can be expressed in mathematical equations for computer analysis, which fits into the testing of theoretical path models.

PATH MODEL DIAGRAMS

Path models adhere to certain common drawing conventions that are utilized in SEM models (Figure 5.1). The observed variables are enclosed by rectangles, and errors of prediction are indicated by oval shapes. Lines drawn from one observed variable to another observed variable denote *direct effects*, that is, the direct influence of one variable on another. A curved arrow between two independent observed variables indicates *covariance*, that is, correlation. This also suggests that such variable relations are influenced by other variables exogenous or external to the path model. Because these influences are not studied in the path model, it is reasonable to expect that some unmeasured variables may influence the independent variables. Each dependent variable in a path model requires an error term, which is denoted by an oval around the error term that points toward the dependent variable. The error term indicates the unexplained error variance of each regression equation.

Model specification is necessary in examining multiple variable relations in path models, just as in the case of multiple regression. Many different relations amongst a set of variables can be hypothesized with many different parameters being estimated. In a simple three-variable model, for example, many possible path models can be postulated on the basis of different hypothesized relations among the three variables. For example, in Figure 5.2, we see three different path models where X1 influences X2. In the first model, X1 influences X2, which in turn influences Y. Here, X2 serves as a mediator between X1 and Y. In the second model, an additional path is drawn from X1 to Y, such that X1 has both a direct and an indirect effect upon Y. The direct effect is that X1 has a direct influence on Y (no variables intervene between X1 and Y), whereas the indirect effect is that X1 influences Y through X2, that is, X2 intervenes between X1 and Y. In the third model, X1 influences both X2 and Y; however, X2 and Y are not related. If we were to switch X1 and X2 around, this would generate three more plausible path models. Other path models are also possible. For example, in Figure 5.3, X1 does not influence X2. In the first model, X1 and X2 influence

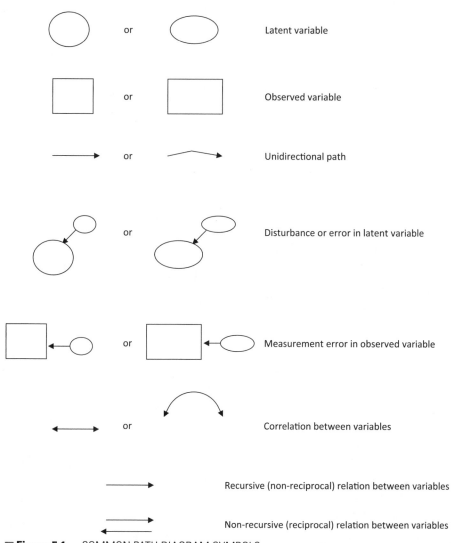

Latent variable

Observed variable

Unidirectional path

Disturbance or error in latent variable

Measurement error in observed variable

Correlation between variables

Recursive (non-reciprocal) relation between variables

Non-recursive (reciprocal) relation between variables

■ **Figure 5.1:** COMMON PATH DIAGRAM SYMBOLS

Y, but are uncorrelated. In the second model, X1 and X2 influence Y and they are correlated.

How can one determine which model is correct? *Model specification* involves using theory and previous research to justify variable relations in a hypothesized theoretical path model. Path analysis does not provide a way to specify the model, but rather estimates the effects of the variables once the model has been specified by the researcher on the basis of theoretical considerations. For this reason, model specification is a critical part of SEM modeling.

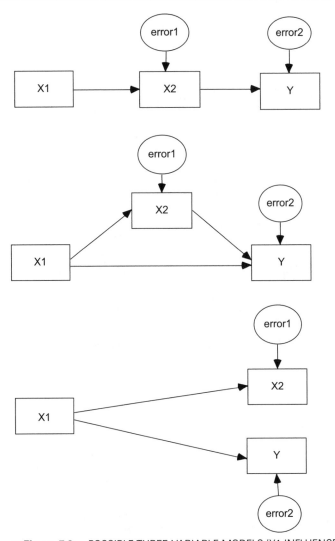

■ Figure 5.2:   POSSIBLE THREE-VARIABLE MODELS (X1 INFLUENCES X2)

Path coefficients in path models are derived from the values of a Pearson product-moment correlation coefficient and/or a standardized partial regression coefficient (Wolfle, 1977). For example, in the second path model of Figure 5.3, the path coefficients (*p*) are depicted by arrows from X1 to Y and X2 to Y, respectively, as

$$\beta_1 = p_{Y1}$$
$$\beta_2 = p_{Y2}$$

and the curved arrow between X1 and X2 is denoted as

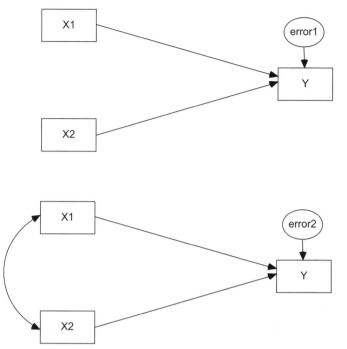

■ **Figure 5.3:**   POSSIBLE THREE-VARIABLE MODELS (X1 DOES NOT INFLUENCE X2)

$$r_{X1,X2} = p_{12}.$$

The variable relations, when specified in standard score form, become standardized partial regression coefficients. In multiple regression, a dependent variable is regressed in a single analysis on all of the independent variables. In path analysis, one or more multiple regression equations are analyzed depending on the variable relations specified in the path model. Path coefficients are therefore computed only on the basis of the particular set of independent variables that lead to the dependent variable under consideration. In the second path model of Figure 5.3, two standardized partial regression coefficients (path coefficients) are computed, p_{Y1} and p_{Y2}. The curved arrow represents the covariance or correlation between the two independent variables p_{12} in predicting the dependent variable.

DECOMPOSITION OF CORRELATION MATRIX

In path models, estimation of parameters permits a decomposition of the correlation matrix. The decomposition of a correlation matrix implies that the original correlation matrix can be completely reproduced if all of the parameters in a path model are specified. For example, take the path model in Figure 5.3 where X1 and

X2 predict Y. There are two direct effects, from X1 to Y and from X2 to Y. There are indirect effects due to the correlation between X1 and X2. X1 indirectly influences Y through X2, and also X2 indirectly affects Y through X1. The correlations among these three variables can be decomposed as follows:

$$r_{12} = p_{12} \tag{1}$$
$$\text{(CO)}$$

$$r_{Y1} = p_{Y1} + p_{12}\,p_{Y2} \tag{2}$$
$$\text{(DE)} \quad \text{(IE)}$$

$$r_{Y2} = p_{Y2} + p_{12}\,p_{Y1}, \tag{3}$$
$$\text{(DE)} \quad \text{(IE)}$$

where the r values are the actual observed correlations and the p values are the path coefficients (standardized estimates).

In equation (1), the correlation between X1 and X2 is simply a function of the path, or correlated relation (CO), between X1 and X2. In equation (2), the correlation between X1 and Y is a function of (a) the direct effect (DE) of X1 on Y, and (b) the indirect effect (IE) of X1 on Y through X2. The indirect effect is computed as the product of the path between X1 and X2 (p_{12}) and the path from X2 to Y (p_{Y2}). Equation (3) is similar to equation (2), except that X1 and X2 are reversed; there is both a direct effect and an indirect effect.

Let us illustrate how this works with an actual set of correlations. The observed correlations are as follows: $r_{12} = .224$, $r_{Y1} = .507$, and $r_{Y2} = .480$. The path coefficients and the complete reproduction of the correlations are:

$$r_{12} = 3\ p_{12} = .224 \tag{4}$$
$$\text{(CO)}$$

$$r_{Y1} = p_{Y1} + p_{12}\,p_{Y2}$$
$$= .421 + (.224)(.386) = .507 \tag{5}$$
$$\text{DE} \qquad\quad \text{IE}$$

$$r_{Y2} = p_{Y2} + p_{12}\,p_{Y1}$$
$$= .386 + (.224)(.421) = .480. \tag{6}$$
$$\text{DE} \qquad\quad \text{IE}$$

The original correlations are completely reproduced because all of the variable relations are indicated in the path model, that is, direct, indirect, and correlated relations.

When a variable relation is left out of a path model, for example, p_{12}, then the correlations would not be completely reproduced. A researcher typically does not specify all the variable relations in a path model based on their theoretical framework. Thus, there will be a difference between the sample correlation matrix (S) and the implied path model correlation matrix (Σ). SEM tests the difference between these two matrices using a chi-square test, $\chi^2 = S - \Sigma$. So the goal is to test a hypothesized path model that does not contain all variable relations, yet closely approximates the variable relations given by the sample variance–covariance matrix. We now understand that if we specified all variable relations in a path model, $\chi^2 = 0$, and the two matrices would be identical. We are not interested in testing hypothesized path models that specify all variable relations. The correlation decomposition is a nice conceptual way of thinking about the model specification process in path analysis, and how much of the correlation amongst variables is explained by a path model. For further details on the correlation decomposition approach, we recommend reading Duncan (1975).

PATH MODEL EXAMPLE

The path model was taken from McDonald and Clelland (1984) who collected data on the sentiments toward unions of Southern non-union textile laborers ($n = 173$). This example was analyzed in the LISREL manual (Jöreskog & Sörbom, 1993, pp. 12–15, example 3); included in the data files with LISREL9; and explained by Bollen (1989, pp. 82–83). The LISREL program is on the book website.

The path model consisted of five observed variables; the independent variables are the number of years worked in the textile mill (actually log of years, denoted simply as years) and worker age (age); the dependent variables are deference to managers (deference), support for labor activism (support), and sentiment toward unions (sentiment). The diagram of the theoretical union sentiment path model is shown in Figure 5.4.

The path model in Figure 5.4 indicates that age and years are correlated (curved arrow). Age has a direct effect on deference and support (straight arrow lines). Deference has a direct effect on support (straight arrow line). Years, deference, and support have a direct effect on sentiment (straight arrow lines). Because deference, support, and sentiment are being predicted, they each have an error term (oval shapes). The path model also shows that six path coefficients will be estimated when analyzing the different regression equations. The correlation between age and years is not estimated. The hypothesized path model does not contain all the variable relations, only the ones based on the researcher's theoretical perspective.

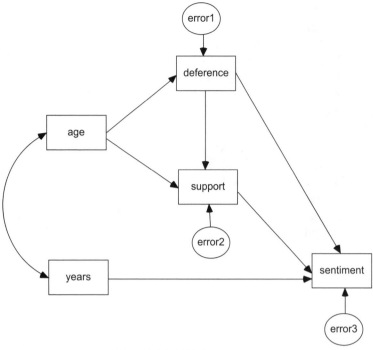

■ **Figure 5.4:**   UNION SENTIMENT MODEL

For the union sentiment path model, there are three regression equations in the model, one for each of the three dependent variables (deference, support, and sentiment). Using the variable names, the regression equations are written as:

deference = (1)age + error1

 support = (2)age + (3)deference + error2

 sentiment = (4)years + (5)support + (6)deference + error3.

SEM software can be used to estimate the six path coefficients for the variables in the regression equations, denoted in parentheses. The software you use will need to specify these three regression equations. Which software you use will be a matter of preference.

Prior research suggested these six relations be included in the path model and that other possible paths, for example, from age to sentiment, not be included. This path model includes direct effects, for example, from age to support; indirect effects, for example, from age to support through deference; and correlated independent variables, for example, age and years. Obviously many possible path models could be specified for this set of observed variables, but theory guided the *model specification* for testing this hypothesized path model.

Model identification is another concern in path models. The *model identification problem* is whether we can obtain estimates for the path coefficients in the model. Basically, given the sample data (sample covariance matrix) and the theoretical path model, can a unique set of parameter estimates be found? In the union sentiment path model, we have chosen to not estimate some relations between variables, that is, parameters. These parameters are fixed. An example of a fixed parameter is when no path or direct relation is specified, for example, between *age* and *sentiment*. The path model has six free parameters, which means they are to be estimated. An example of a free parameter is the path or direct relation for *age* predicting *deference*.

In model identification, we consider the *order condition*. The *order condition* specifies that the number of free parameters to be estimated must be less than or equal to the number of distinct values in the sample variance–covariance matrix. In the union sentiment path model, we have the following free parameter values:

6 path coefficients
3 equation error variances (for 3 dependent variables)
1 correlation among the independent variables
2 independent variable variances.

There are 12 free parameters in the path model. The number of distinct values in the sample variance–covariance matrix is computed by

$$[p(p + 1)]/2 = [5(5 + 1)]/2 = 15,$$

where p is the number of observed variables in the matrix. The number of distinct values in the sample variance–covariance matrix, 15, is greater than the number of free parameters, 12, in the path model. The order condition is satisfied.

The *degrees of freedom* (*df*) for a path model is the difference between the number of distinct values and the number of free parameters. The $df = 15 - 12 = 3$ for this path model. The degrees of freedom should be equal to or greater than 1 for model identification. Therefore, the order condition has been met, and we have a positive degrees of freedom. The path model is an *over-identified* model because there are more distinct values in the sample variance–covariance matrix than free parameters to be estimated in the path model. In SEM, we prefer over-identified models, rather than saturated just-identified regression models, where $\chi^2 = 0$ and df = 0.

The order condition is easy to check by the researcher; however, the *rank condition* of the sample variance–covariance matrix is more challenging. The rank of a matrix is defined as the maximum number of linearly independent rows or columns.

The rank of a matrix can also be interpreted as sufficient when the determinant of a matrix is non-zero. Path models use matrix algebra to compute parameter estimates, thus certain matrix conditions must be present, otherwise parameters can't be estimated. Many of the SEM software programs return the determinant of the sample variance–covariance matrix, or provide condition codes and error messages when there are problems with the matrices. In LISREL9, the following values were returned, which indicated a positive determinant, positive range of eigenvalues, but a condition code that indicated multicollinearity.

```
Total Variance = 372.694 Generalized Variance = 788.073
Largest Eigenvalue = 341.373 Smallest Eigenvalue = 0.074
Condition Number = 67.976
WARNING: The Condition Number indicates severe multicollinearity.
One or more variables may be redundant.
```

Model estimation is choosing the estimation method to estimate the parameters in the path model. When data meet parametric statistical assumptions, the unweighted least squares estimation method works fine. Many SEM software programs default to the maximum likelihood (ML) estimation method, but this is not always a wise choice, especially when data are not normally distributed. The choice of an estimation method can affect whether a parameter estimate is statistically significant, that is, the value of the standard error estimate may be inflated or biased. High-speed computers today provide the opportunity to select each of the estimation methods available and determine their effect on the values and significance of the parameter estimates. User guides for SEM software can guide your selection of an estimation method given the level of measurement of each variable, non-normality, and type of model (Muthén & Muthén, 2006, pp. 423–426).

In the union sentiment path model, the maximum likelihood estimation method was used to obtain the parameter estimates. The path model parameters in Table 5.1 were all statistically significantly at $p < .05$. If any path coefficients were non-significant, we would drop the path from the equation and re-run the path analysis. We accepted the condition of multicollinearity amongst the variables.

Age had a direct effect on both *deference* and *support*; *deference* had a direct effect on *sentiment*; *years* had a direct effect on *sentiment*; and *support* had a direct effect on *sentiment*. We would interpret the path coefficients similar to regression coefficients, so the positive and negative coefficients are meaningful. For example, *age* predicting *deference* has a negative path coefficient, so younger age workers had less deference for union management than older workers. *Support* and number of *years* working in the textile mill had positive path coefficients in predicting *sentiment* toward unions. This implies that more support for labor activism and longer years of working at the textile mill are related to increased sentiment for unions. Indirect effects, although indicated in the path model, are not presented in the table. The indirect effects, called moderator effects, are obtained by multiplying the standardized path coefficients. For example, *deference* is a moderator variable between *age* and *sentiment*, so −.336 times −.147 yields .049 path coefficient. The standardized path coefficients, direct and indirect effects can be obtained using the command: LISREL OUTPUT SC EF; the standardized path coefficients are reported in Table 5.1.

The three regression equations, which reported the unstandardized path coefficients, standard errors, z values, and p values were:

```
Structural Equations
Deferenc =  - 0.0874*Age, Errorvar.= 12.961, R² = 0.113
Standerr     (0.0187)            (1.402)
Z-values     -4.664              9.246
P-values     0.000               0.000


 Support =  - 0.285*Deferenc + 0.0579*Age, Errorvar.= 8.488,
                                                     R² = 0.230
Standerr     (0.0619)            (0.0161)         (0.918)
Z-values     -4.598              3.597            9.246
P-values     0.000               0.000            0.000


Sentimen =  - 0.218*Deferenc + 0.850*Support + 0.861*Years,
                               Errorvar.= 19.454, R² = 0.390
Standerr     (0.0974)    (0.112)        (0.341)         (2.104)
Z-values     -2.235      7.555          2.526           9.246
P-values     0.025       0.000          0.012           0.000
```

The *model fit* was assessed by a chi-square test (Table 5.1). Other model-fit indices are also available to indicate whether the path model reflects the sample variance–covariance relations amongst the observed variables. If the sample data are a good fit to the theoretical hypothesized path model, then the chi-square will be close to zero, and non-significant. This indicates that the sample variance–covariance matrix is similar to the path model-implied variance–covariance matrix. The $\chi^2 = 1.26$, $df = 3$, $p = .74$, so the small non-significant chi-square value indicated that the path model reflects the observed variable relations in the sample variance–covariance matrix. If the hypothesized path model was not a good representation of the sample data, then the chi-square statistic would be statistically significant.

Table 5.1: Union Sentiment Path Model

Paths	Standardized Parameter Estimates	Unstandardized Parameter Estimates	Standard Error	Z	p
Age → deference	−.336	−.087	.018	−4.66	<.001
Age → support	.256	.058	.016	3.59	<.001
Deference → support	−.328	−.285	.061	−4.59	<.001
Years → sentiment	.154	.86	.341	2.52	.01
Deference → sentiment	−.147	−.218	.097	−2.23	.02
Support → sentiment	.499	.850	.112	7.55	<.001
Equation error variances					
Deference		12.96			
Support		8.49			
Sentiment		19.45			
Model fit					
χ^2		1.26			
df		3			
p value		.74			

SEM software prints out the sample variance–covariance matrix (*S*), the path model-implied variance–covariance matrix (Σ), and a residual matrix. These are reported in Table 5.2. The difference between the elements of the sample variance–covariance (*S*) and the model-implied variance–covariance matrix (Σ) produces the residual matrix. A non-significant chi-square indicates that the values in the residual matrix would be small, that is, not much difference between the two matrices. When chi-square is statistically significant, large residuals should be present that indicate where the variable relations are problematic. Standardized residuals, *z* scores greater than 1.96 or 2.58, would indicate that a particular variable relation is not well accounted for in the path model. The standardized residual values in Table 5.2 are small as expected, and do not exceed $z = 1.96$.

Table 5.2: Original, Model-implied, Residual, and Standardized Residual Covariance Matrices for Union Sentiment Path Model

Original Covariance Matrix

Variable	Deference	Support	Sentiment	Years	Age
Deference	14.610				
Support	−5.250	11.017			
Sentiment	−8.057	11.087	31.971		
Years	−0.482	0.677	1.559	1.021	
Age	−18.857	17.861	28.250	7.139	215.662

Table 5.2: (*cont.*)

Model-implied Covariance Matrix

Variable	Deference	Support	Sentiment	Years	Age
Deference	14.610				
Support	−5.250	11.017			
Sentiment	−8.179	11.013	31.899		
Years	−0.624	0.591	1.517	1.021	
Age	−18.857	17.861	25.427	7.139	215.662

Residual Covariance Matrix

Variable	Deference	Support	Sentiment	Years	Age
Deference	0.000				
Support	0.122	0.000			
Sentiment	0.142	0.074	0.072		
Years	0.142	0.086	0.042	0.000	
Age	0.000	0.000	2.823	0.000	0.000

Standardized Residual Covariance Matrix

Variable	Deference	Support	Sentiment	Years	Age
Deference	0.000				
Support	0.000	0.000			
Sentiment	0.495	0.338	0.167		
Years	0.586	0.412	0.207	0.000	
Age	0.000	0.000	0.702	0.000	0.000

SUMMARY

Path models provide an understanding of how to diagram and specify a model, estimate parameters, and test model fit. The decomposition of a correlation matrix provides a keen insight into how SEM modeling works. Basically, a hypothesized model is specified without all variable relations included and run with sample data to produce parameter estimates. If the model is a good representation of the variable relations, then the parameter estimates produce a similar variance–covariance matrix. A chi-square test of the difference between the sample matrix and model-implied matrix, yields a test of model fit. We desire a non-significant chi-square value to indicate that the two matrices are similar.

The union sentiment path model example provided a better understanding of model specification, model estimation, and model fit. It showed that several regression equations are included in a path model, and that path coefficients are interpreted similar to regression coefficients. It further provided an opportunity to discuss issues related to path analysis, for example, choice of estimation method and whether parameters could be estimated based on the determinant of the matrix and eigenvalues.

Path models permit specification of theoretically meaningful relations amongst variables that cannot be specified in a single additive regression model. Path models provide the additional ability to test indirect effects. Indirect effects provide for a test of moderator variables. The issue of measurement error in observed variables, however, is not treated in either regression or path models (Wolfle, 1979).

EXERCISE

Analyze the following achievement path model (Figure 5.5). The path model indicates that income and ability predict aspire, and income, ability, and aspire predict achieve. Sample size = 100. The variance–covariance matrix is in Table 5.3.

Observed variables:

```
quantitative achievement (Ach)
family income (Inc)
quantitative ability (Abl)
educational aspiration (Asp)
```

Table 5.3: Variance–covariance matrix ($n = 100$)

	Ach	Inc	Abl	Asp
Ach	25.500			
Inc	20.500	38.100		
Abl	22.480	24.200	42.750	
Asp	16.275	13.600	13.500	17.000

Equations:

Asp = Inc + Abl

Ach = Inc + Abl + Asp

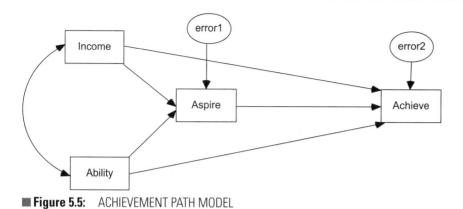

■ **Figure 5.5:** ACHIEVEMENT PATH MODEL

CHAPTER FOOTNOTE

Traditional Path Model Fit

The comparison of the original and reproduced correlation matrices is essential for testing the significance of the path model (Specht, 1975). SEM software provides a chi-square test of matrix difference. However, the traditional path analysis model fit was given by R^2 to indicate how much variance in the dependent variable was explained.

For the union sentiment path model, three regression equations were computed, thus three R^2 values are reported. These three R^2 values are used to compute the overall R^2_m for the path model. This is accomplished by computing the generalized squared multiple correlation (Pedhazur, 1982) as follows:

$$R^2_m = 1 - (1 - R^2_1)(1 - R^2_2)...(1 - R^2_p).$$

The R^2 values are the squared multiple correlation coefficients from each of the separate regression analyses in the path model. The $1 - R^2$ is the unexplained variance of each regression equation. The path model-fit R^2_m would be computed as

$$
\begin{aligned}
R^2_m &= 1 - (1 - .113)(1 - .230)(1 - .39) \\
&= 1 - (.887)(.77)(.61) \\
&= 1 - .42 \\
&= .58
\end{aligned}
$$

other types of correlation coefficients. The different factor analysis techniques were applied because of nominal and ordinal measurement. For example, test questions scored 0 (incorrect) and 1 (correct) were viewed as different from questionnaire item responses coded strongly agree (SA), agree (A), neutral (N), disagree (D), and strongly disagree (SD). Exploratory factor analysis was used to determine the number of factors needed to explain the correlations among a set of observed variables. EFA was considered a data reduction method, where the factors were fewer than the number of variables, and required the factors to be subjectively named based on which items shared common variance. The common variance or commonality was indicated by the strength of the factor loading. In EFA, the level of measurement of the variables can be binary, ordinal, or continuous, but requires the use of different correlations and estimation methods. Mplus and LISREL, for example, both provide EFA methods under different scoring conditions. A common mistake made by researchers is that they conveniently use the wrong EFA technique, in part because they are unaware that different factor analysis methods are available, and that the different EFA methods are based on the level of measurement of the variables. So, in summary, there are different types of exploratory factor analysis methods available in SEM.

Exploratory factor analysis is considered an indeterminate solution because more than one configuration of variables exists that could produce the factors. Consequently, there are several issues in exploratory factor analysis that researchers generally address in practice (Comrey & Lee, 1992; Costello & Osborne, 2005; Gorsuch, 1983). These are summarized next.

Sample size: Basically, it takes only two pairs of scores to compute a correlation coefficient. Would the correlation coefficient be a good sample estimate of the population correlation, rho? The issues become one of sampling, inference, and validity. The more stable the sample correlations, the more valid the inference of the scores to the population. Costello and Osborne (2005) reviewed journal articles and conducted simulations to recommend best practices in exploratory factor analysis. They suggested that 20 subjects per variable result in reasonable solutions.

Number of factors: The determination of how many factors has been done using a scree plot with the number of eigenvalues plotted against the number of factors (Cattell, 1966). The rule of thumb has been to select the number of factors based on eigenvalues greater than one. A correlation (variance–covariance matrix) contains one or more eigenvalues, and they appear in decreasing order of magnitude. So, the first factor will have the largest eigenvalue (largest variance explained), the second factor a lower eigenvalue (lesser variance explained), and so forth. The exploratory nature is to find the fewest factors that have the largest amount of variance accounted for, and provide a meaningful interpretation (Schumacker, 2015).

Rotation methods: Factor analysis is invariant within rotations, that is, the initial factor pattern matrix is not unique. We can get an infinite number of solutions, which produce the same correlation matrix, by rotating the reference axes of the factor solution to simplify the factor structure and to achieve a more meaningful and interpretable solution. Rotation methods provide a different orientation of data points to factors. This also points out the indeterminate nature of exploratory factor analysis.

Factor scores: Exploratory factory analysis permits the computation of factor scores. Factor scores are usually computed using a multiple regression equation where the factor loadings are multiplied by their respective variables. Factor scores are output as standardized scores, but can be converted to a scaled score using a linear transformation. There are three methods used to compute factor scores: regression (Thurstone, 1935); Bartlett (Bartlett, 1937), and Anderson–Rubin (Anderson & Rubin, 1956). The regression method maximizes the validity of the construct (factor variance explained). The Bartlett method computes factor scores keeping the factors orthogonal (uncorrelated). The Anderson–Rubin method computes factor scores that have correlations which match the correlations amongst the factors, thus only computed with oblique factors. SEM software uses the regression method where the standardized factor loadings are used as regression weights to compute a score.

EFA vs PCA: Exploratory factor analysis (EFA) is often confused with principal components analysis (PCA). Both methods extract variance from the correlation matrix. Factor analysis extracts variable variance based on the squared multiple correlation in the diagonal of the matrix. Principal components analysis extracts variable variance, but from the diagonal of the correlation matrix where each variable has a variance = 1.0, so the sum of the diagonal is the number of variables in the correlation matrix. PCA components reproduce all of the variable variance. Factor analysis is designed to account for only the variance in the partial correlation of each variable, so it does not include all the variable variance. Principal components analysis is designed to account for all of the variable variance, as indicated by the 1's in the diagonal of the matrix. This distinction is not always made clear, thus principal components is sometimes analyzed, when in fact the researcher intended to conduct a factor analysis (Schumacker, 2015).

Factor analysis attempts to determine which sets of observed variables share common variance–covariance characteristics that define theoretical constructs or factors (latent variables). Factor analysis presumes that some factors that are smaller in number than the number of observed variables are responsible for the shared variance–covariance amongst the observed variables. In practice, one collects data on observed variables and uses factor-analytic techniques to either *confirm* that a particular subset of observed variables defines each construct or factor, or *explore* which observed variables relate to factors. In exploratory factor model approaches, we seek to find a model that fits the data, so we specify different alternative models, hoping to ultimately find a model that fits the data and has theoretical support. This is the primary rationale for exploratory factor analysis (EFA). In confirmatory factor model approaches, we seek to statistically test the significance of a hypothesized factor model, that is, whether the sample data confirm that model. Additional samples of data that fit the model further confirm the validity of the hypothesized model. This is the primary rationale for confirmatory factor analysis (CFA).

In CFA, the researcher specifies a certain number of factors, which factors are correlated, and which observed variables measure each factor. In EFA, the researcher explores how many factors there are, whether the factors are correlated, and which observed variables appear to best measure each factor. In CFA, the researcher has an *a priori* specified theoretical model; in EFA, the researcher does not have such a model. Examples of EFA and CFA are given below.

EFA EXAMPLE

There are numerous sample data sets available on the internet (www.inside-r. org/howto/finding-data-internet), which can be used for statistical analyses. I am using the 11 subscales on the Weschler Intelligence Scale for Children from a sample of 175 learning disabled children (Tabachnick & Fidell, 2007). The 11 subscales measure the following: info (Information), comp (Comprehension), arith (Arithmetic), simil (Similarities), vocab (Vocabulary), digit (Digit Span), pictcomp (Picture Completion), parang (Paragraph Arrangement), block (Block Design), object (Object Assembly), and coding (Coding). There are 16 subjects per variable (175 children divided by 11 variables), and no missing data. The internet link, data set, and program are on the book website.

Mplus software was used to input the raw data from the ASCII text file, *WISC.txt*, which is read by the *data* command. The variables are listed on the *variable* command. The *client* number and *agemate* variables were excluded using the *usevar* command. The *analysis* command specifies an EFA analysis with a minimum of one factor and a maximum of three factors. The *plot* command produces the scree plot. There are other commands for missing data and rotation method, which are not included. The Mplus EFA program by default uses the unweighted least squares estimation method. The Mplus program syntax is:

```
title: Exploratory factor analysis
data: file is "c:\WISC.txt";
variable: names are client agemate info comp arith simil
      vocab digit pictcomp parang block
      object coding;
usevar are info comp arith simil
      vocab digit pictcomp parang block
      object coding;
analysis: type = EFA 1 3;
plot: type = plot2;
```

The abbreviated Mplus output indicates that the program executed properly, the estimation method was unweighted least squares (ULS), and lists the name of the input data file.

```
INPUT READING TERMINATED NORMALLY
Exploratory factor analysis
Estimator                    ULS
Input data file(s)
 c:\WISC.txt
```

The program output lists the eigenvalues for each of the 11 variables; we want this range of variable eigenvalues to have positive values. The criteria for selecting the number of factors is based on the number of eigenvalues greater than 1. There are 3 eigenvalues greater than 1, so we can reduce the 11 variables into 3 factors. The 3 factors should represent enough variance for all 11 variables in the sample variance–covariance matrix. This is why exploratory factor analysis is considered a data reduction method.

```
EIGENVALUES FOR SAMPLE CORRELATION MATRIX
    1      2      3      4      5      6      7      8
 3.829  1.442  1.116  0.890  0.768  0.633  0.595  0.522

    9     10     11
 0.471  0.419  0.315
```

The eigenvalues of the factors are generally given by the matrix formula:

$$E = V'RV$$

where E = eigenvalues, V = eigenvectors or weights, and R = correlation matrix (Schumacker, 2015). Eigenvalues indicate the amount of variance associated with each variable in the correlation (covariance) matrix. The eigenvalues are extracted from a Wishart distribution, $Wp(n;§)$, where p is the dimension, n is the degrees of freedom and § is the covariance matrix. Anderson (1963) provided the large sample theory of eigenvalues and eigenvectors, where the distribution of the sample eigenvalues is a chi-square distribution with associated degrees of freedom. We are interested in eigenvalues because they indicate the amount of variance that can be explained by each variable in a given correlation (covariance) matrix, and yield associated eigenvector weights for each variable on the factors.

The scree plot in Figure 6.1 provides a visual representation of the eigenvalues by the number of variables. The eigenvalues are on the Y axis and the number of variables on the X axis.

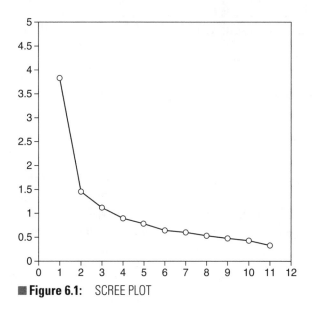

■ **Figure 6.1:**   SCREE PLOT

The 3-factor solution is examined based on 3 eigenvalues greater than 1.0, which should explain the relations amongst all 11 variables. The output performs a varimax rotation to provide a simpler solution. A variable is determined to be on a factor with the highest factor loading, which is based on comparing their factor loadings in the *structure* matrix. For example, INFO has the highest factor loading on Factor 1. I have boldfaced the highest factor loadings for each variable in Table 6.1. It appears that the first factor contains INFO, COMP, ARITH, SIMIL, VOCAB, and DIGIT. The second factor contains PICTCOMP, PARANG, BLOCK, and OBJECT. The third factor contains a single item, CODING.

The factor solution is not desirable because a single observed variable loaded on the third factor. We will therefore drop the CODING variable, and re-run the EFA with a two-factor solution.

Table 6.1: EFA Varimax Rotated Factor Loadings

	VARIMAX ROTATED LOADINGS		
	1	2	3
INFO	**0.783**	0.147	0.020
COMP	**0.562**	0.432	−0.025
ARITH	**0.556**	0.145	0.206
SIMIL	**0.628**	0.353	−0.155
VOCAB	**0.721**	0.225	0.020
DIGIT	**0.426**	0.011	0.242
PICTCOMP	0.217	**0.587**	−0.209
PARANG	0.164	**0.406**	0.124
BLOCK	0.154	**0.676**	0.219
OBJECT	0.078	**0.591**	−0.023
CODING	0.051	0.030	**0.431**

```
title: Exploratory factor analysis drop variable - coding
data: file is "c:\WISC.txt";
variable: names are client agemate info comp arith simil
     vocab digit pictcomp parang block
     object coding;
usevar are info comp arith simil
     vocab digit pictcomp parang block
     object;
analysis: type = EFA 2 2;
```

The factor structure matrix now indicates a clear bi-factor solution, as shown in Table 6.2. The observed variable loadings for each factor selected are boldfaced.

The two factors are correlated $r = .50$. The chi-square model fit was non-significant ($\chi^2 = 33.92$, $df = 26$, $p = .137$), so the EFA measurement model fit the sample variance–covariance matrix.

The researcher is now faced with naming these factors. The first factor might be called *verbal ability*, and the second factor *spatial ability*. If we used the factor loadings in a regression equation, we could compute factor scores for these two latent traits, and then use the factor scores in other statistical analyses. Croon (2002) provides cautionary remarks about using predicted latent scores as observed variables because it leads to biased estimates of the joint distribution of a latent variable with observed variables, and also yields inconsistent estimates of the joint distribution of two or more latent variables. Another rationale is that factor analysis provides an indeterminate solution, so the

Table 6.2: EFA Bi-factor Solution

| | FACTOR STRUCTURE | |
	1	2
INFO	0.803	0.370
COMP	0.636	0.594
ARITH	0.580	0.294
SIMIL	0.665	0.525
VOCAB	0.767	0.436
DIGIT	0.419	0.134
PICTCOMP	0.326	0.626
PARANG	0.246	0.431
BLOCK	0.319	0.637
OBJECT	0.211	0.603

factor scores can easily change. The factor scores using the regression method can be computed as follows:

```
verbal = .803(INFO) + .636(COMP) + .580(ARITH) + .665(SIMIL) + .767(VOCAB) + .419(DIGIT)

spatial = .626(PICTCOMP) + .431(PARANG) + .637(BLOCK) + .603(OBJECT)
```

PATTERN AND STRUCTURE MATRICES

In a single-factor solution, the pattern and structure matrices would be identical. In a bi-factor solution, they would be different. The pattern matrix holds the factor loadings. Each row of the *pattern* matrix is essentially a regression equation where the standardized observed variable is expressed as a function of the factors. The loadings are the regression coefficients. The *structure* matrix holds the correlations between the variables and the factors.

The pattern matrix has factor loadings similar to partial standardized regression coefficients, which indicates the effect of a given factor for a specific variable controlling for other variables. The structure matrix is the zero-order Pearson correlation of the variable with the factor, which helps in interpreting which variable goes with which factor for interpretation. When two or more factors are present, these two matrices differ. The more correlated the factors, the greater the difference between the factor pattern and structure loadings. When the factors are orthogonal, the factor pattern and structure matrices are identical. The factor pattern matrix is used when obtaining factor scores, and reproducing the correlation matrix (Pett, Lackey, & Sullivan, 2003). The factor structure matrix is Pφ, or pattern matrix times the covariance (correlation) matrix. When factors are orthogonal, φ = I (identity matrix), so PI = P, and the pattern and structure matrices are identical.

CONFIRMATORY FACTOR ANALYSIS

The validity and reliability issues in measurement have traditionally been handled by first examining the validity and reliability of scores on instruments used in a particular design context. Given an acceptable level of score validity and reliability, the scores are then used in a statistical analysis. However, the traditional statistical analysis of these scores, for example, in multiple regression and path analysis, does not adjust for measurement error. The impact of measurement error has been investigated and found to have serious consequences, for example, it yields biased parameter estimates (Cochran, 1968; Fuller, 1987). A general approach to confirmatory factor analysis using maximum likelihood estimation was therefore developed by Jöreskog (1969).

Confirmatory factor analysis (CFA) tests a hypothesized theoretical measurement model, that is, determines whether the hypothesized measurement model yields a variance–covariance matrix similar to the sample variance–covariance matrix. In CFA, a researcher specifies which variables go together, and are assigned to a factor, thus yielding a *pattern* matrix. It is considered bad practice to first conduct an exploratory factor analysis, then use the results in a confirmatory factor analysis. You would obviously obtain your desired results. The *pattern* matrix is used in CFA with specific variables indicated for their respective factors or latent variables, which corresponds to the diagram of the theoretical measurement model. Variables with a factor loading set to zero on any factor would indicate no relation, so a path would not appear in the measurement model diagram. Also, rather than use a rule of thumb for the value of a factor loading, say .60 or higher, the significance of the factor loading should be used to determine which observed variables are important (Cudeck & O'Dell, 1994).

CFA EXAMPLE

Holzinger and Swineford (1939) collected data on 26 psychological tests from seventh- and eighth-grade children in a suburban school district of Chicago. Over the years, different subsamples of the children and different subsets of the variables of this data set have been analyzed and presented in various multivariate statistics textbooks, for example, Harmon (1976) and Gorsuch (1983), and SEM user guides, for example, Jöreskog and Sörbom (1993; example 5, pp. 23–28).

The raw data analyzed here are on the first 6 psychological variables for all 301 subjects; the resulting sample covariance matrix (*S*) is given in the SIMPLIS program. The confirmatory factor model consists of the following six observed variables: *Visual Perception, Cubes, Lozenges, Paragraph Comprehension, Sentence Completion,* and *Word Meaning.* The first three measures were hypothesized to measure a *spatial* ability factor and the second three measures to measure a *verbal* ability factor.

The path diagram of the theoretical proposed model is shown in Figure 6.2. The observed variables are enclosed by boxes or rectangles, and the factors (latent variables) are enclosed by circles, that is, *spatial* and *verbal.* Conceptually, a factor represents the common variation among a set of observed variables. Thus, for example, the *spatial* ability factor represents the common variation among the *Visual Perception, Cubes,* and *Lozenges* tasks. Lines directed from a factor to a particular observed variable denote the relations between that factor and that measure. These relations are interpreted as factor loadings with the square of the factor loading called the *commonality estimate* of the variable.

The measurement errors are enclosed by smaller ovals and indicate that some portion of each observed variable is measuring something other than the hypothesized factor. Conceptually, a measurement error represents the unique variation for a particular observed variable beyond the variation due to the relevant factor. For example, the *Cubes* task is largely a measure of *spatial* ability, but may also be assessing other characteristics such as a different common factor or unreliability. To assess measurement error, the variance of each measurement error is estimated (known as *measurement error variance*).

A curved arrow between two factors indicates that they have shared variance or are correlated. In this example, *spatial* and *verbal* ability are specified to covary or correlate. The rationale for this particular factor correlation is that *spatial* ability and *verbal* ability are related to a more general ability factor and thus should be theoretically related.

A curved arrow between two measurement error variances indicates that they also have shared variance or are correlated. Although not shown in this example, two measurement error variances could be correlated if they shared something in common such as (a) common method variance where the method of measurement is the same, such as the same scale of measurement, or they are both part of the same global instrument, or (b) the same measure is being used at different points in time, that is, the *Cubes* task is measured at Time 1 and again at Time 2.

The *model specification* is a first step in confirmatory factor analysis, just as it was for multiple regression and path models. Model specification is necessary because many different relations amongst a set of variables can be postulated with many different parameters being estimated. Thus, many different factor models can be postulated on the basis of different hypothesized relationships between the observed variables and the factors.

Our CFA model is specified to have six observed variables with two different latent variables (factors) being hypothesized. In Figure 6.2, each observed variable is hypothesized to measure only a single factor, that is, three observed variables per factor with six factor loadings; the factors are believed to be correlated (a single factor correlation); and the measurement error variances are not related (zero correlated measurement errors). How does the researcher determine which factor model is correct? We already know that *model specification* is important in this process and indicates the important role that theory and prior research play in justifying a specified model. Confirmatory factor analysis does not tell us how to specify the model, but rather estimates the parameters of the model once the model has been specified by the researcher. Model specification is the hardest part of structural equation modeling.

The CFA model in Figure 6.2 contains six measurement equations in the model, one for each of the six observed variables. In terms of the variable names from Figure 6.2, the measurement equations are as follows:

$$\text{visperc} = \lambda_{11} \text{ spatial} + \text{err\_v}$$
$$\text{cubes} = \lambda_{12} \text{ spatial} + \text{err\_c}$$

$$\text{lozenges} = \lambda_{13} \text{ spatial} + \text{err\_l}$$

$$\text{paragrap} = \lambda_{21} \text{ verbal} + \text{err\_p}$$
$$\text{sentence} = \lambda_{22} \text{ verbal} + \text{err\_s}$$
$$\text{wordmean} = \lambda_{23} \text{ verbal} + \text{err\_w}$$

Substantive theory and prior research suggest that factor loadings (λ_{11} to λ_{23}) should be estimated for these variables in the specified model, and that other possible factor loadings, for example, *visperc* loading on *verbal*, should not be included in the confirmatory factor model. Our CFA model includes six factor loadings and six measurement error variances, one for each observed variable, and one correlation between the factors *spatial* ability and *verbal* ability with zero correlated measurement errors.

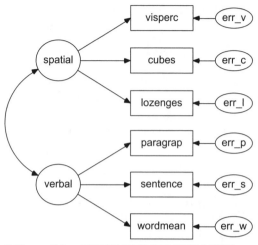

■ **Figure 6.2:**   CONFIRMATORY FACTOR MODEL

Model identification is the next step in the confirmatory factor analysis. It is crucial that the researcher check for any *identification problem* prior to the estimation of parameters. For our confirmatory factor model, we would like to know if the factor loading of *Visual Perception* on *Spatial Ability*, *Cubes* on *Spatial Ability*, *Lozenges* on *Spatial Ability*, *Paragraph Comprehension* on *Verbal Ability*, *Sentence Completion* on *Verbal Ability*, and *Word Meaning* on *Verbal Ability* can be estimated. In our confirmatory factor model, some parameters are fixed and others are free. An example of a *fixed parameter* is that *Cubes* is not allowed to load on *Verbal Ability*. An example of a *free parameter* is that *Cubes* is allowed to load on *Spatial Ability*.

In determining identification, we first assess the *order condition*. The number of free parameters to be estimated must be less than or equal to the number of distinct values in the matrix *S*. A count of the free parameters is as follows:

6 factor loadings
6 measurement error variances
0 measurement error covariance terms or correlations
1 correlation among the latent variables.

Thus, there are a total of 13 free parameters that we wish to estimate. The number of distinct values in the matrix *S* is equal to

$$p(p + 1)/2 = 6(6 + 1)/2 = 21,$$

where *p* is the number of observed variables in the sample variance–covariance matrix. The number of values in *S*, 21, is greater than the number of free parameters, 13, with the difference being the degrees of freedom for the specified model, *df* = 21 − 13 = 8. However, this is only a necessary condition and does not guarantee that the model is identified. According to the order condition, this model is *over-identified* because there are more values in S than parameters to be estimated, that is, our degrees of freedom is positive not zero (*just-identified*) or negative (*under-identified*). We seek over-identified models in SEM. We have also discovered the role a positive determinant of the sample variance–covariance matrix plays in our ability to extract variance (eigenvalues) and obtain parameter estimates.

The next step is to estimate the factor loadings for the hypothesized factor model. In factor analysis the traditional method of estimation is to decompose the variance–covariance matrix. The complete decomposition of the variance–covariance matrix is possible when all variable relations are accounted for by the factors in a specified factor model, reflected in the structure matrix. When a factor model is hypothesized with fewer factors, the variance–covariance matrix will not be completely reproduced, reflected in the pattern matrix. The goal is to have a hypothesized factor model that reproduces most of the original sample variance–covariance matrix.

The factor loadings can be estimated by different estimation procedures, such as maximum likelihood (ML), generalized least squares (GLS), and unweighted least squares (ULS), and reported as unstandardized estimates or standardized estimates. We analyzed our confirmatory factor model using maximum likelihood estimation with a standardized solution to report our statistical estimates of the factor model parameters. We used the SIMPLIS program to specify the CFA model as follows:

```
LISREL-SIMPLIS Confirmatory Factor Model Program
Confirmatory Factor Model Figure 6.2
Observed Variables:
VISPERC CUBES LOZENGES PARCOMP SENCOMP WORDMEAN
Sample Size: 301
Covariance Matrix
        49.064
        9.810 22.182
        27.928 14.482 81.863
        9.117 2.515 5.013 12.196
        10.610 3.389 3.605 13.217 26.645
        19.166 6.954 13.716 18.868 28.502 58.817
Latent Variables: Spatial Verbal
Relationships:
VISPERC CUBES LOZENGES = Spatial
PARCOMP SENCOMP WORDMEAN = Verbal
Print Residuals
Number of Decimals = 3
Path Diagram
End of problem
```

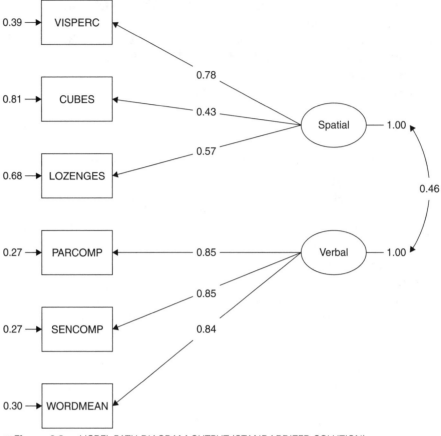

▪ **Figure 6.3:** LISREL PATH DIAGRAM OUTPUT (STANDARDIZED SOLUTION)

The program syntax is upper- and lower-case sensitive, so capitalized variable names must also be used in the *Relationships* commands, which specifies which observed variables go with the latent variables. The latent variables are named using the *Latent Variable* command. The *Print Residual* command will output the residual values (difference between elements of the S and Σ matrices).

The software provides a computer output file and a graph of the CFA model. The CFA model was drawn in LISREL, where the standardized solution was chosen from the pull-down menu (Figure 6.3).

The standardized factor pattern loadings, factor correlation of *spatial* and *verbal*, and chi-square model fit are reported in Table 6.3. The chi-square model-fit value was statistically significant, so the sample variance–covariance matrix was not a good fit to the CFA measurement model.

Table 6.3: CFA Factor Model

Factor Model	Spatial	Verbal
Factor loadings:		
Visual Perception	.78	0.00
Cubes	.43	0.00
Lozenges	.57	0.00
Paragraph Comprehension	0.00	.85
Sentence Completion	0.00	.85
Word Meaning	0.00	.84
(Spatial, Verbal)	.461	
χ^2	24.36	
df	8	
p value	.002	

When the CFA model is not a good fit to the sample variance–covariance matrix, the researcher has several options to choose from. First, the standardized values in the residual matrix could be examined. If any values are greater than $z = 1.96$, they could indicate a possible additional path in the model. Second, some SEM software programs provide guidance by printing out modification indices (MI). MIs indicate where paths could be added or deleted to improve model fit. Third, the structure matrix could be computed to determine if the CFA model was correctly specified to indicate observed variable relations on factors (Graham, Guthrie & Thompson, 2003).

It is possible to obtain the structure coefficient matrix from the product of the Lambda X pattern matrix times the Phi matrix: $\psi_s = \Lambda_x \Phi$. You can use R, SPSS, SAS, or other software to multiply the matrices. For the CFA results in Table 6.3, R was used to calculate the structure coefficients as follows:

$$\Psi_s = \Lambda_x \Phi$$

$$\Psi_s = \begin{bmatrix} .78 & 0 \\ .43 & 0 \\ .57 & 0 \\ 0 & .85 \\ 0 & .85 \\ 0 & .84 \end{bmatrix} \begin{bmatrix} 1 & .461 \\ .461 & 1 \end{bmatrix}$$

$$\Psi_s = \begin{bmatrix} .78 & .36 \\ .43 & .20 \\ .57 & .26 \\ .39 & .85 \\ .39 & .85 \\ .39 & .84 \end{bmatrix}$$

The structure coefficients (Ψ_s) indicated that the pattern coefficients (Λ_x) were specified correctly, that is, the observed variables were indicators of factors where they had the highest factor loading. We therefore turn our attention to the MIs, especially where adding a path for error covariance makes theoretical sense. The SIMPLIS program provides MIs when a model does not fit. The CFA model showed the following output:

```
The Modification Indices Suggest to Add an Error Covariance
Between        and         Decrease in Chi-Square   New Estimate
LOZENGES       CUBES             171.7                  71.76
PARCOMP        VISPERC           436.7                  95.55
PARCOMP        CUBES              13.1                   4.71
PARCOMP        LOZENGES           23.2                  15.43
SENCOMP        VISPERC           280.6                 115.25
SENCOMP        CUBES              11.0                   6.37
WORDMEAN       VISPERC           316.4                 158.78
WORDMEAN       CUBES              20.3                  12.70
WORDMEAN       LOZENGES           34.0                  39.79
```

A researcher would generally pick the MI with the largest value. However, not all of these MIs will make sense to use given our theoretical CFA model. For example, we would not want to add error covariance between PARCOMP and VISPERC because they fall on different factors. Also, they reflect different types of observed variables, one being paragraph completion and the other visual perception, so theoretically it would not make sense. The MIs for WORDMEAN and VISPERC, as well as SENCOMP and VISPERC, have a similar concern. We therefore look at LOZENGES and CUBES, which are both related and on the same factor, *Spatial*. We substantiate the added error covariance between these observed variables because they shared a common method variance where the method of measurement was the same (same scale of measurement), and they were both part of the same instrumentation in the study.

To modify the program with this error covariance, we add the following command line after our CFA model statements on the *Relationships* command:

```
Let the error covariances between LOZENGES and CUBES correlate
```

This resulted in our final solution, reported in Table 6.4.

The chi-square model-fit value was non-significant ($\chi^2 = 13.68$, $df = 7$, $p = .06$), which indicated a good data to model fit. The added error covariance term decreased the model degrees of freedom by one, so $df = 7$. The path diagram now

Table 6.4: Modified CFA Factor Model

Factor Model	Spatial	Verbal
Factor Loadings:		
Visual Perception	.96	0.00
Cubes	.31	0.00
Lozenges	.46	0.00
Paragraph Comprehension	0.00	.85
Sentence Completion	0.00	.86
Word Meaning	0.00	.84
(Cubes, Lozenges)	.20	
(Spatial, Verbal)	.42	
χ^2	13.68	
df	7	
p value	.06	

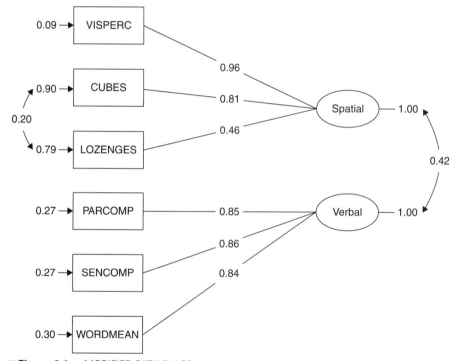

■ Figure 6.4: MODIFIED PATH DIAGRAM

includes a curved arrow between CUBES and LOZENGES, which indicates the coefficient for the error covariance. The curved arrow should appear in the path diagram, as shown in Figure 6.4.

SUMMARY

There are many different SEM software packages to choose from. In this chapter, I used Mplus and LISREL for EFA and CFA, respectively. A researcher could have used AMOS, EQS, SAS-Proc Calis, or any number of other software packages. SEM software today is easy to use, requires only a few command lines, and produces a substantial amount of computer output and graphics for testing SEM models.

EFA is used when a researcher has no prior theoretical basis for a measurement model. It also implies that the model is not being tested for fit, that is, whether the model is a good representation of the sample variance–covariance matrix. The goal of EFA is data reduction where a few factors explain the variable relations. A researcher determines which variables go with which factors by selecting those with the highest factor loadings in the structure matrix. A second sample of data can be obtained, and the EFA factor structure can be tested using CFA methods.

CFA is for testing a hypothesized theoretical model. A researcher, based on theory and prior research, specifies which variables load on which factors. Consequently, observed variables have fixed parameters of zero on factors they do not represent. The pattern matrix therefore has fixed zero values to indicate a variable does not load on that factor. Modification indices are provided in many software packages to guide adding or deleting paths in the model tested, when a good data to model fit is not indicated.

There are a few options a researcher can use when modifying a CFA model. The standardized residual matrix, the MIs, and the structure coefficients. Generally, adding an error covariance term will improve the model fit, thus yielding a non-significant chi-square value. Justification for the added error covariance should be given. Substantial changes would require a new theoretical support for the CFA model tested. The relation between the pattern matrix and factor correlations was given to compute the structure coefficients. Structure coefficients are more prominently interpreted in EFA models, but can provide clear evidence of a misspecified CFA model when an observed variable shows a higher loading on a different factor than indicated in the CFA model.

Factor analysis provides the measurement models where a set of observed variables define a factor. Multiple measures of a construct are considered to yield more valid and reliable scores than a single measure. The individual factor score (latent variable score) indicates the composite of these multiple observed variable measures, when computed using the regression method. CFA in SEM forms the basis for creating a specified measurement model that yields latent scores used in the structural model, which tests relations amongst latent variables. The various SEM models tested in the following chapters will serve to illustrate the diverse applications.

EXERCISE

Test the hypothesized confirmatory factor model shown in Figure 6.5. Use the correlation matrix in Table 6.5 with a sample size of 3094. The observed variables are:

```
Academic ability (Academic)
Self-concept (Concept)
Degree aspirations (Aspire)
Degree (Degree)
Occupational prestige (Prestige)
Income (Income)
```

The CFA model indicates that the first three observed variables measure the latent variable Academic Motivation (Motivate) and the last three observed variables measure the latent variable Socioeconomic Status (SES). Motivate and SES are correlated.

Table 6.5: Correlation Matrix (n = 3094)

Academic	Concept	Aspire	Degree	Prestige	Income
1.000					
0.487	1.000				
0.236	0.206	1.000			
0.242	0.179	0.253	1.000		
0.163	0.090	0.125	0.481	1.000	
0.064	0.040	0.025	0.106	0.136	1.000

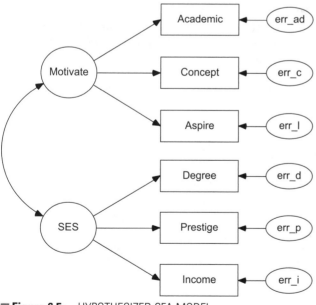

■ **Figure 6.5:**   HYPOTHESIZED CFA MODEL

REFERENCES

Anderson, R. D., & Rubin, H. (1956). Statistical inference in factor analysis. *Proceedings of the Third Berkeley Symposium of Mathematical Statistics and Probability*, 5, 111–150.

Anderson, T. W. (1963). Asymptotic theory for principal component analysis. *Annals of Mathematical Statistics*, 36, 413–432.

Bartlett, M. S. (1937). The statistical conception of mental factors. *British Journal of Psychology*, 28, 97–104.

Cattell, R. B. (1966). The scree test for the number of factors. *Multivariate Behavioral Research*, 1(2), 245–276.

Cochran, W. G. (1968). Errors of measurement in statistics. *Technometrics*, 10, 637–666.

Comrey, A. L., & Lee, H. B. (1992). *A first course in factor analysis*. Hillsdale, NJ: Lawrence Erlbaum.

Costello, A. B., & Osborne, J. (2005). Best practices in exploratory factor analysis: Four recommendations for getting the most from your analysis. *Practical Assessment Research and Evaluation*, 10(7), 1–9.

Croon, M. (2002). Using predicted latent scores in general latent structure models. In Marcoulides, G. A., & Moustaki, I. (Eds.), *Latent Variable and Latent Structure Models*. Mahwah, NJ: Lawrence Erlbaum Associates.

Cudeck, R., & O'Dell, L. L. (1994). Applications of standard error estimates in unrestricted factor analysis: Significance tests for factor loadings and correlations. *Psychological Bulletin*, 115, 475–487.

Esposito Vinzi, V., Chin, W. W., Henseler, J., & Wang, H. (2010). *Handbook of partial least squares: Concepts, methods, and applications* (Eds). New York: Springer Verlag.

Fuller, W. A. (1987). *Measurement error models*. New York: Wiley.

Gorsuch, R. L. (1983). *Factor analysis* (2nd edn.). Hillsdale, NJ: Lawrence Erlbaum.

Graham, J. M., Guthrie, A. C., & Thompson, B. (2003). Consequences of not interpreting structure coefficients in published CFA research: A reminder. *Structural Equation Modeling*, 10, 142–153.

Harmon, H. H. (1976). *Modern factor analysis* (3rd edn, rev.). Chicago: University of Chicago Press.

Holzinger, K. J., & Swineford, F. A. (1939). *A study in factor analysis: The stability of a bi-factor solution*. (Supplementary Educational Monographs, No. 48.) Chicago, IL: University of Chicago, Department of Education.

Jöreskog, K. G. (1969). A general approach to confirmatory maximum likelihood factor analysis. *Psychometrika*, 34, 183–202.

Jöreskog, K. G., & Sörbom, D. (1993). *LISREL 8: Structural equation modeling with the SIMPLIS command language*. Chicago, IL: Scientific Software International.

Pett, M. A., Lackey, N. R., & Sullivan, J. J. (2003). *Making sense of factor analysis: The use of factor analysis for instrument development in health care research*. Thousand Oaks, CA: Sage Publications.

Schumacker, R. E. (2015). *Using R with multivariate statistics.* Thousand Oaks, CA: Sage Publications.

Tabachnick, B. G., & Fidell, L. S. (2007). *Using multivariate statistics* (5th edn.). New York: Pearson Education.

Thurstone, L. L. (1935). *The vectors of mind* (pp. 226–231). Chicago: University of Chicago Press.

Wold, H. (1966). Estimation of principal components and related models by iterative least squares. In Krishnaiaah, P. R. (Ed.), *Multivariate analysis* (pp. 391–420). New York: Academic Press.

Wold, H. (1985). Partial least squares. In Kotz, S., & Johnson, N. L. (Eds.). *Encyclopedia of statistical sciences, 6* (pp. 581–591). New York: Wiley.

Chapter 7

SEM BASICS

CHAPTER CONCEPTS

SEM Modeling Steps
   Model specification
   Model identification
   Model estimation
   Model testing
   Model modification
Model-fit Criteria
   Saturated models and independence models
   Measurement versus structural models
Types of Fit Criteria
   Model fit
   Model comparison
   Model parsimony
   Parameter fit
Hypothesis Testing
   Power
   Sample size
   Model comparison
   Effect size
Parameter Significance
Two-Step Versus Four-Step Approach to Modeling

Structural equation modeling (SEM) tests whether a sample variance–covariance matrix is similar to the variance–covariance matrix implied by a theoretical model. The theoretical model produces a variance–covariance matrix based on the paths specified in the model (recall path models). The chi-square statistic tests whether the two matrices are similar or different. A significant chi-square value indicates that the two matrices are different, thus the sample data do not support the theoretical model, while a non-significant chi-square value indicates that the theoretical model produces a matrix similar to the sample data matrix. This has been referred to as a "badness of fit" test because we seek a non-significant chi-square value.

There are several approaches when conducting SEM to test whether a sample variance–covariance matrix is supported by a theoretical model. For example, a CFA *measurement model* could be tested to determine if the variables share common variance in defining a latent variable. A researcher is attempting to establish key variables that relate to a construct. The theoretical model is either confirmed or disconfirmed, based on the chi-square statistical test of significance and/or meeting acceptable model-fit criteria. Another SEM approach could specify *alternative models* where the researcher creates different theoretical models. The theoretical models use the same data set, so they are referred to as *nested* models. The alternative approach conducts a chi-square difference test to compare each of the alternative models. In this situation, a significant chi-square value would indicate that the two models being compared were different. A third SEM approach is considered *model generating*, where an initial theoretical model is specified, but when the data do not fit this initial model *modification indices* are used to add or delete paths in the model to arrive at a final best-fitting model. The goal in model generating is to find a model that the data fit well with a practical and substantive theoretical meaning. The process of finding the best-fitting model is also referred to as a *specification search*, implying that if an initially specified model does not fit the data, then the model is modified in an effort to improve the fit (Marcoulides & Drezner, 2001, 2003). Recent advances in *Tabu search algorithms* have permitted the generation of a set of models that the data fit equally well with a final determination by the researcher of which model to accept (Marcoulides, Drezner, & Schumacker, 1998). AMOS includes a specification search procedure to find the best-fitting model.

SEM MODELING STEPS

The steps a researcher takes in conducting SEM involves *model specification, model identification, model estimation, model testing,* and *model modification.* SEM combines measurement models (CFA) with structural models (path models) using latent variables. Once the measurement models for both latent independent and dependent variables yield a good data to model fit, the relations amongst the latent variables are tested in the structural model. A researcher spends most of their time selecting observed variables as indicators of latent variables, thus testing the CFA models. The structural model tests the parameter estimates in the structural equations for statistical significance. The SEM modeling steps generally occur as outlined in the following five basic steps:

Model specification: A measurement model and/or a structural model are specified based on prior research and theory. This comprises the review of literature which substantiates selection of observed variables as indicators of latent variables, and the theory behind testing the relations amongst the latent variables in a structural model.

Model identification: A model is identified if the degrees of freedom is equal to or greater than 1. A $df = 0$ indicates a saturated model, thus all parameters are being estimated, which is also called a just-identified model. An under-identified model would have negative degrees of freedom because more parameters are being estimated than distinct values in the covariance matrix. We are interested in an over-identified model, which specifies fewer paths or variable relations, yet the model-implied (reproduced) variance matrix is close to the sample covariance matrix. The order and rank condition are also other matrix features which indicate a model is identified, which implies that parameters can be estimated.

Model estimation: A hypothesized theoretical model can have parameters estimated using several different estimation methods. The unweighted least squares estimation method works fine when the assumptions of the Pearson correlation coefficient are met: normal distribution assumption, and other parametric assumptions hold. Other estimation methods, such as maximum likelihood, were developed to handle data that do not meet the parametric assumptions, but yield robust estimates of parameters. A key issue in estimating model parameters is the associated standard error. If the standard error is biased or inflated, then the test of statistical significance of a parameter is affected. A model parameter is tested for statistical significance by dividing the parameter estimate by its standard error.

Model testing: A model is tested for fit based on the non-significance of the chi-square statistic, and other subjective indices. When chi-square is non-significant it indicates that the original variance–covariance matrix and the model-implied variance–covariance matrix are similar. This implies that the model is a good representation of the relations amongst the observed variables, that is, their variance and covariance.

Model modification: When hypothesis testing, a model may not fit the data. A researcher is guided by residual values in the residual matrix, modification indices, or theory to make changes. It is not recommended that the structural model be changed by adding or deleting a path unless additional theory substantiates the structural model modification. Generally, the measurement model will require adding an error covariance term between observed variables, which is sufficient to provide a better data to model fit. When an error covariance term is added, justification should be provided, which usually includes similar observed variables on a factor, the same measurement scale, or similar instrumentation. It is also plausible that the model simply does not fit the data, which implies that the theory is not supported or perhaps another random sample of data will work better.

MODEL-FIT CRITERIA

Structural equation modeling (SEM) tests a hypothesized theoretical model, which has its basis in testing theory. Theory is defined as relations amongst constructs. Constructs are first established by testing confirmatory factor models, which create latent variables. The latent variables are then used in a structural equation model, which forms the basis of the theory.

A researcher typically uses the following three criteria in judging the statistical significance and substantive meaning of a theoretical model:

1. The first criterion is the non-statistical significance of the chi-square test, which is considered a global fit measure. A non-statistically significant chi-square value indicates that the sample covariance matrix and the reproduced model-implied covariance matrix are similar.
2. The second criterion is the statistical significance of individual parameter estimates for the paths in the model, which are values computed by dividing the parameter estimates by their respective standard errors. This is referred to as a t or z value, and is typically compared to a tabled t or z value of 1.96 at the .05 level of significance.
3. The third criterion is the magnitude and direction of the parameter estimates, paying particular attention to whether a positive or negative coefficient makes sense for the parameter estimate. For example, it would not be theoretically meaningful to have a negative parameter estimate for number of hours spent studying and grade point average.

There are numerous subjective model-fit criteria for assessing model fit. Determining model fit is complicated because several model-fit criteria have been developed to assist in interpreting structural equation models under different model-building assumptions. In addition, the determination of model fit in structural equation modeling is not as straightforward as it is in other statistical approaches in multivariable procedures, such as the analysis of variance, multiple regression, discriminant analysis, multivariate analysis of variance, and canonical correlation analysis. These multivariable methods use observed variables that are assumed to be measured without error and have statistical tests with known distributions. Many SEM model-fit indices have no single statistical test of significance that identifies a correct model, given the sample data, especially as *equivalent models* or *alternative models* can exist that yield exactly the same data to model fit (Hershberger & Marcoulides, 2013).

Chi-square (χ^2) is considered the only statistical test of significance for testing the theoretical model. The chi-square value ranges from zero for a saturated model with all paths included to a maximum value for the independence model with no paths included. Your theoretical model chi-square value will be somewhere between these two extreme values. This can be visualized as follows:

Saturated model (all paths in model) $\chi^2 = 0$	$\Leftrightarrow$	Independence model (no paths in model) $\chi^2 = $ maximum value

A chi-square value of zero indicates a perfect fit or no difference between values in the sample covariance matrix (S) and the model-implied covariance matrix (Σ) that was created, based on the specified theoretical model. Obviously, a theoretical model in SEM with all paths specified is of limited interest because it always yields a saturated model (recall regression equations). The goal in structural equation modeling is to achieve a parsimonious model with a few substantive meaningful paths and a non-significant chi-square value close to the saturated model value of zero, thus indicating little difference between the sample covariance matrix and the model-implied covariance matrix. The difference between these two covariance matrices is output in a residual matrix. When the chi-square value is non-significant (close to zero), residual values in the residual matrix are close to zero, indicating that the sample data fit the theoretical implied model; hence there is little difference between the sample covariance matrix and the model-implied covariance matrix.

Many of the model-fit criteria are computed-based on knowledge of the saturated model, independence model, sample size, degrees of freedom, and/or the chi-square values, which are used to formulate an index of model fit that ranges in value from 0 (no fit) to 1 (perfect fit). These various model-fit indices, however, are subjectively interpreted when determining an acceptable model fit. Some researchers have suggested that a structural equation model with a model-fit value of .95 or higher is acceptable (Baldwin, 1989; Bentler & Bonett, 1980), while others suggested the non-centrality parameter close to zero (Browne & Cudeck, 1993; Steiger, 1990) and the root-mean-square error of approximation (Steiger, 1990). Consequently, the various structural equation modeling programs report a variety of model-fit criteria. We distinguish the model-fit criteria based on whether assessing *model fit, model comparison,* or *model parsimony* as global fit measures (Hair, Anderson, Tatham, & Black, 1992).

Many of the subjective fit indices are computed given knowledge of the null model χ^2 (independence model, where the covariance terms are assumed to be zero in the model), null model *df,* hypothesized model χ^2, hypothesized model *df,* number of observed variables in the model, number of free parameters in the model, and sample size. The formula for the goodness-of-fit index (GFI), normed fit index (NFI), relative fit index (RFI), incremental fit index (IFI), Tucker–Lewis index (TLI), comparative fit index (CFI), model AIC, null AIC, and RMSEA using these values are as follows:

```
GFI = 1 - [χ²model/χ²null]
NFI = (χ²null - χ²model)/χ²null
RFI = 1 - [(χ²model/dfmodel)/(χ²null/dfnull)]
IFI = (χ²null - χ²model)/(χ²null - dfmodel)
TLI = [(χ²null/dfnull) - (χ²model/dfmodel)]/[(χ²null/dfnull) - 1]
CFI = 1 - [(χ²model - dfmodel)/(χ²null - dfnull)]
Model AIC = Normal theory χ²model + 2q (Note: q = df -1)
Null AIC = χ²null + 2q (Note: q = df - 1)
```

$$RMSEA = \sqrt{\chi^2_{Model} - df_{Model} / [(N-1)df_{Model}]}$$

These model-fit statistics can also be expressed in terms of the non-centrality parameter (NCP), designated by γ. The estimate of NCP (γ_{Model}) using the maximum likelihood chi-square is $\chi^2 - df$, that is, $NCP = (\chi^2 - df) = \gamma_{Model}$. A simple substitution expresses these model-fit statistics using NCP. For example, CFI, TLI, and RMSEA are as follows:

```
CFI = 1 - [γModel/γNull]
TLI = 1 - [(γModel/dfModel)/(γNull/dfNull)]
```

$$RMSEA = \sqrt{\gamma_{Model} / [(N-1)df_{Model}]}$$

Bollen and Long (1993), as well as Hu and Bentler (1995), have thoroughly discussed several issues related to model fit, and we recommend reading their assessments of how model-fit indices are affected by small sample bias, estimation methods, violation of normality and independence, and model complexity, and for an overall discussion of the various model-fit indices. We have summarized some of these model-fit criteria in Table 7.1.

Table 7.1: Model-fit Criteria and Acceptable Fit Interpretation

Model-fit Criterion	Acceptable Level	Interpretation
Chi-square	Tabled χ^2 value	Compares obtained χ^2 value with tabled value for given *df*
Goodness-of-fit index (GFI)	0 (no fit) to 1 (perfect fit)	Value close to .90 or .95 reflects a good fit
Adjusted GFI (AGFI)	0 (no fit) to 1 (perfect fit)	Value adjusted for *df*, with .90 or .95 a good model fit
Root-mean square residual (RMR)	Researcher defines level	Indicates the closeness of Σ to S matrices
Standardized RMR (SRMR)	< .05	Value less than .05 indicates a good model fit
Root-mean-square error of approximation (RMSEA)	.05 to .08	Value of .05 to .08 indicates close fit
Tucker–Lewis Index (TLI)	0 (no fit) to 1 (perfect fit)	Value close to .90 or .95 reflects a good model fit
Normed fit index (NFI)	0 (no fit) to 1 (perfect fit)	Value close to .90 or .95 reflects a good model fit
Parsimony fit index (PNFI)	0 (no fit) to 1 (perfect fit)	Compares values in alternative models
Akaike information criterion (AIC)	0 (perfect fit) to positive value (poor fit)	Compares values in alternative models

TYPES OF MODEL-FIT CRITERIA

Overall, we placed the fit indices into the three main categories of *model fit*, *model comparison*, and *model parsimony*. Extensive comparisons and discussions of many of these fit indices can be found in issues of the following journals: *Structural Equation Modeling: A Multidisciplinary Journal*, *Psychological Bulletin*, *Psychological Methods*, and *Multivariate Behavioral Research*.

Model Fit

Model fit determines the degree to which the sample variance–covariance data fit the structural equation model. Model-fit criteria commonly used are chi-square (χ^2), the goodness-of-fit index (GFI), the adjusted goodness-of-fit index (AGFI), the root-mean-square error of approximation (RMSEA), and the root-mean-square residual index (RMR) (Jöreskog & Sörbom, 1993). These criteria are based on differences between the observed (original, S) and model-implied (reproduced, Σ) variance–covariance matrices.

Chi-square (χ^2)

A significant χ^2 value relative to the degrees of freedom indicates that the observed and implied variance–covariance matrices are different. Statistical significance indicates the probability that this difference is due to sampling variation. A non-significant χ^2 value indicates that the two matrices are similar, indicating that the implied theoretical model significantly reproduces the sample variance–covariance relations in the matrix. The researcher is interested in obtaining a non-significant χ^2 value with associated degrees of freedom. Thus it may be more appropriate to call the chi-square test a measure of badness-of-fit.

The chi-square test of model fit can lead to erroneous conclusions regarding analysis outcomes. The χ^2 model-fit criterion is sensitive to sample size because as sample size increases (generally above 200), the χ^2 statistic has a tendency to indicate a significant probability level. In contrast, as sample size decreases (generally below 100), the χ^2 statistic indicates non-significant probability levels. The chi-square statistic is therefore affected by sample size, as noted by its calculation, $\chi^2 = (n - 1)F_{ML}$, where F is the maximum likelihood (ML) fit function. The χ^2 statistic is also sensitive to departures from multivariate normality of the observed variables.

Three estimation methods are commonly used to calculate χ^2 in latent variable models (Loehlin, 1987): maximum likelihood (ML), generalized least squares (GLS), and unweighted least squares (ULS). Each approach estimates a best-fitting solution and evaluates the model fit. The ML estimates are consistent, unbiased, efficient, scale-invariant, scale-free, and normally distributed if the observed variables meet the multivariate normality assumption. The GLS estimates have the same properties as the ML approach under a less stringent multivariate normality assumption and provide an approximate chi-square test of model fit to the data. The ULS estimates do not depend on a normality distribution assumption; however, the estimates are not as efficient, nor are they scale-invariant or scale-free. The ML χ^2 statistic is $\chi^2 = (n - 1)F_{ML}$, the GLS χ^2 statistic is $\chi^2 = (n - 1)F_{GLS}$, and the ULS χ^2 statistic is $\chi^2 = (n - 1)F_{ULS}$. Check your SEM software user guide to determine what estimation methods are available, when to use them, and what different chi-square values are reported (see Chapter Footnote for LISREL).

Goodness-of-fit Index (GFI) and Adjusted Goodness-of-fit Index (AGFI)

The goodness-of-fit index (GFI) is based on the ratio of the sum of the squared differences between the observed and reproduced matrices to the observed variances, thus allowing for scale. The GFI measures the amount of variance and covariance in S that is predicted by the reproduced matrix, Σ. If GFI = .97, then

97% of the S matrix is predicted by the reproduced matrix Σ. The GFI index can be computed for ML, GLS, or ULS estimates (Bollen, 1989). The basic formula is

$$\text{GFI} = 1 - [\chi^2_{model}/\chi^2_{null}]$$

(*Note*: χ^2_{null} is the Chi-Square for Independence Model value reported on the computer output.)

The adjusted goodness-of-fit index (AGFI) is adjusted for the degrees of freedom of a model relative to the number of variables. The AGFI index is computed as $1 - [(k/df)(1 - \text{GFI})]$, where k is the number of unique distinct values in S, which is $p(p+1)/2$, and df is the number of degrees of freedom in the model. For GFI = .985, $k = 15$ distinct values in the S matrix, and $df = 7$, the AGFI index is:

```
1 - [(k/df)(1 - GFI)] = 1 - [(15/7)(1 - .985)]
= 1 - [2.14285(.015)]
= 1 - .03
= .97
```

The GFI and AGFI indices can be used to compare the fit of two different alternative models with the same data, or compare the fit of a single model using different data, such as separate data sets for males and females in a multiple group model.

Root-mean-square Error of Approximation (RMSEA)
The RMSEA provides a measure of close fit when the computed value is between .05 and .08. The RMSEA is computed as

$$\text{RMSEA} = \sqrt{\chi^2_{\text{Model}} - df_{\text{Model}} / [(N-1)df_{\text{Model}}]}$$

RMSEA takes into account the model degrees of freedom and sample size. The degrees of freedom is considered a measure of model complexity. The rationale is that for more degrees of freedom, there is an increase in the number of variable relations in the model. Similarly, the relation between sample size and degrees of freedom is incorporated into the RMSEA value, which is standardized.

Root-mean-square Residual Index (RMR)
The RMR index uses the square root of the mean-squared differences between matrix elements in S and Σ. Because it has no defined acceptable level, it is best used to compare the fit of two different alternative models with the same data. The RMR index is computed as

$$\text{RMR} = [(1/k) \, \Sigma_{ij} \, (s_{ij} - \Sigma_{ij})^2]^{1/2}$$

There is also a standardized RMR, known as Standardized RMR, which has an acceptable level when less than .05.

Model Comparison

A comparison of alternative models uses the following three model-fit indices: the Tucker–Lewis index (TLI) or Bentler–Bonett non-normed fit index (NNFI), the Bentler–Bonett normed fit index (NFI) (Bentler & Bonett, 1980; Loehlin, 1987), and the comparative fit index (CFI). These three criteria typically compare a proposed model with a null model (independence model). The null model is indicated by the independence-model chi-square value. The null model could also be any model that establishes a baseline from which one could expect other alternative models to be different.

Tucker–Lewis Index (TLI)

Tucker and Lewis (1973) initially developed the TLI for factor analysis but later extended it to structural equation modeling. The measure can be used to compare alternative models or to compare a proposed model against a null model. The TLI is computed using the χ^2 statistic as

$$[(\chi^2_{null} / df_{null}) - (\chi^2_{proposed} / df_{proposed})]/[(\chi^2_{null} / df_{null}) - 1]$$

It is scaled from 0 (no fit) to 1 (perfect fit). For our modified model analysis, the NNFI is computed as

$$\text{NNFI} = [(\chi^2_{null} / df_{null}) - (\chi^2_{proposed} / df_{proposed})]/[(\chi^2_{null} / df_{null}) - 1]$$

Normed Fit Index (NFI) and Comparative Fit Index (CFI)

The NFI is a measure that rescales chi-square into a 0 (no fit) to 1.0 (perfect fit) range (Bentler & Bonett, 1980). It is used to compare a restricted model with a full model using a baseline null model as follows: $(\chi^2_{null} - \chi^2_{model})/\chi^2_{null}$. The NFI is computed as

$$\text{NFI} = (\chi^2_{null} - \chi^2_{model})/\chi^2_{null}$$

Bentler (1990) subsequently developed a coefficient of comparative fit within the context of specifying a population parameter and distribution, such as a population comparative fit index, to overcome the deficiencies in NFI for nested models. The rationale for assessment of comparative fit in the nested-model approach involves a series of models that range from least restrictive (M_i) to saturated (M_s).

Corresponding to this sequence of nested models is a sequence of model-fit statistics with associated degrees of freedom. The comparative fit index (CFI) measures the improvement in non-centrality in going from model M_i to M_k (the theoretical model) and uses the non-central χ^2 (d_k) distribution with non-centrality parameter γ_k to define comparative fit as $(\gamma_i - \gamma_k)/\gamma_i$.

McDonald and Marsh (1990) further explored the non-centrality and model-fit issue by examining nine fit indices as functions of non-centrality and sample size. They concluded that only the Tucker–Lewis index and their relative non-centrality index (RNI) were unbiased in finite samples and recommended them for testing null or alternative models. For absolute measures of fit that do not test null or alternative models, they recommended d_k (Steiger & Lind, 1980), because it is a linear function of χ^2, or a normed measure of centrality m_k (McDonald, 1989), because neither of these varied systematically with sample size. These model-fit measures of centrality are useful when selecting among a few competing models based upon theoretical considerations.

Model Parsimony

Parsimony refers to the number of estimated parameters required to achieve a specific level of model fit. Basically, an over-identified model is compared with a restricted model. The AGFI measure discussed previously also provides an index of model parsimony. Other indices that indicate model parsimony are the parsimony normed fit index (PNFI), and the Akaike information criterion (AIC). Parsimony-based fit indices for multiple indicator models were reviewed by Williams and Holahan (1994). They found that the AIC performed the best (see their article for more details on additional indices and related references). The model parsimony goodness-of-fit indices take into account the number of parameters required to achieve a given value for chi-square. Lower values for PNFI and AIC indicate a better model fit given a specified number of parameters in a model.

Parsimony Normed Fit Index (PNFI)

The PNFI measure is a modification of the NFI measure (James, Mulaik, & Brett, 1982). The PNFI, however, takes into account the number of degrees of freedom used to obtain a given level of fit. Parsimony is achieved with a high degree of fit for fewer degrees of freedom in specifying the coefficients to be estimated. The PNFI is used to compare models with different degrees of freedom and is calculated as PNFI = $(df_{\text{proposed}} / df_{\text{null}})$ NFI.

Akaike Information Criterion (AIC)

The AIC measure is used to compare models with differing numbers of latent variables, much as the PNFI is used (Akaike, 1987). The AIC can be calculated in two different ways: $\chi^2 = 2q$, where q = number of free parameters in the model, or as $\chi^2 - 2df$. The first AIC is positive, and the second AIC is negative, but either AIC value close to zero indicates a more parsimonious model. The AIC indicates model fit (S and Σ elements similar) and model parsimony (over-identified model). AIC uses normal theory weighted least squares chi-square not the minimum fit function chi-square. SEM software will generally indicate which AIC is reported.

Model AIC = Normal Theory $\chi^2 + 2q$

Independence AIC = Chi-Square for Independence Model + $2(df - 1)$

Summary

Mulaik, James, Alstine, Bennett, Lind, and Stilwell (1989) evaluated the χ^2, NFI, GFI, AGFI, and AIC goodness-of-fit indices. They concluded that these indices fail to assess parsimony and are insensitive to misspecification of structural relationships (see their definitive work for additional information). Their findings should not be surprising because it has been suggested that a *good* fit index is one that is independent of sample size, accurately reflects differences in fit, imposes a penalty for inclusion of additional parameters (Marsh, Balla, & McDonald, 1988), and supports the choice of the true model when it is known (McDonald & Marsh, 1990). *No model-fit criteria can actually meet all of these criteria.*

There are several model-fit indices because scholars could not agree on how model fit, model comparison, and model parsimony should be reported. Researchers generally report more than one model-fit criterion. SEM software user guides also report and explain these model-fit criteria. We recommend that once you feel comfortable using these fit indices for your specific model applications, you check the references cited for additional information on their usefulness and/ or limitations. The model-fit criteria have stirred much controversy and discussion on their subjective interpretation and appropriateness under specific modeling conditions (see Marsh, Balla, & Hau, 1996 for further discussion). Further research and discussion will surely follow; for example, Kenny and McCoach (2003) indicated that RMSEA improves as more variables are added to a model, whereas TLI and CFI both decline in correctly specified models as more variables are added.

When deciding on which model-fit indices to report, first consider whether the fit indices were created for model fit, model parsimony, or model comparison. At the risk of oversimplification, we suggest that χ^2, RMSEA, and Standardized RMR be reported for most types of models with additional fit indices reported based on the type of model tested. For example, the CFI should be reported if comparing alternative models. Overall, more than one model-fit index should be reported. If

(no difference in matrices) against an alternative model (difference in matrices). Hypothesis testing involves understanding power. Power in traditional parametric tests is the ability to detect a difference, and thus reject the null hypothesis and accept the alternative hypothesis. In SEM, we have to reframe our thinking of power as the ability to retain the null hypothesis and reject the alternative hypothesis. Power and related issues (sample size, degrees of freedom, alpha, directional nature of hypothesis) impact our decision of whether sample data fit a theoretical model. Power is therefore discussed next.

Power

The determination of power and/or sample size in SEM is complicated because theoretical models can have several variables or parameter estimates, and parameters are typically not independent in a model, and can have different standard errors. In SEM we also compare models, oftentimes nested models with the same data set. Consequently, power and sample size determination in the situation where a researcher is hypothesis testing (testing a model fit to data), comparing alternative models, or desiring to test a parameter estimate for significance, will be covered with SAS, SPSS, G*Power 3, R examples. The *power* for hypothesis testing, or the probability of retaining H_o when H_a is false, depends on the true population model, significance level, degrees of freedom, and sample size, which involves specifying an effect size, alpha, and sample size; while *sample size* determination is achieved given power, effect size, and alpha level of significance.

The *significance* of parameter estimates that do not require two separate models to yield separate χ^2 values includes: (a) generating a two-sided t or z value for the parameter estimate (T or Z = parameter estimate divided by standard error of the parameter estimate), and (b) interpreting the modification index directly for the parameter estimate as a χ^2 test with 1 degree of freedom. The relationship is simply $T^2 = D^2 = $ MI (modification index) for large sample sizes. Gonzalez and Griffin (2001), however, indicated that the standard errors of the parameter estimates are sensitive to how the model is identified, that is, alternative ways of identifying a model may yield different standard errors, and hence different T values for the statistical significance of a parameter estimate. This lack of invariance due to model identification could result in different conclusions about a parameter's significance level from different, yet equivalent, models on the same data. The authors recommended that parameter estimates be tested for significance using the likelihood ratio (LR) test because it is invariant to model identification, rather than the t test (or z test).

Saris and Satorra (1993) provided an easy to use approach for calculating the power of a theoretical model. Basically, an alternative model is estimated with sample data to indicate what percent of the time we would correctly reject the null hypothesis under the assumption that the null hypothesis (H_o) is false. The minimum fit function chi-square value obtained from fitting data to the theoretical model provides an estimate of the non-centrality parameter (NCP; λ). NCP is calculated as $\chi^2 - df_{model}$ given normal theory weighted least squares estimation. This makes computing power using SAS, SPSS, or G*Power 3 straightforward using their respective command functions. We are using $\chi^2 = 3.841$, $df = 1$, $p < .05$ as the critical tabled value for testing our hypothesis of model fit, given the comparison between two alternative models. However, when testing a single model fit, the model chi-square and degrees of freedom would be used. Power examples for NCP = 6.63496 are provided next to show SAS and SPSS syntax.

SAS syntax—power

```
data chapter7;
do obs=1;
ncp = 6.63496;
power = 1 - PROBCHI(3.841, 1, ncp);
output;
end;
proc print;
var ncp power;
run;
```

SPSS syntax—power

```
DATA LIST FREE / obs.
BEGIN DATA.
1
END DATA.
compute ncp = 6.63496.
compute power = 1 - NCDF.CHISQ(3.841, 1, ncp).
formats ncp power (f8.5).
List.
```

Given NCP = 6.63496, power = .73; the output from the SAS or SPSS syntax was:

obs	ncp	power
1.00	6.63496	.73105

Power, given the *ncp* value from a single model, can also be determined using G*power 3 (Faul, Erdfelder, Lang & Buchner, 2007). Power and sample size estimates for a-priori and post-hoc statistical applications are available using G*power 3. In the *Test family* window select "χ^2 tests"; in the Statistical test window select "Generic χ^2 test"; and in the *Type of power analysis* window, select "*Post-hoc: Compute power – given* α, *and non-centrality parameter.*" Click the *Calculate* button. The power = .731015 value should appear. We interpret the power value as a 73% chance of retaining the null hypothesis at the .05 level of significance, which falls short of the .80 level commonly accepted for power. If we replace the critical chi-square value in the formula, you can determine the power for other alpha levels of significance. In Table 7.2, we have replaced the critical chi-square value and ran the SPSS syntax program for alpha values ranging from .10 to .001. If we test model fit at $p = .10$ level, then we achieve an acceptable level of power; other alpha levels from .05 to .001 fall below a .80 power value.

Table 7.2: Power for Alpha Levels of NCP

NCP	Critical Chi-square	Alpha	Power
6.63496	2.706	.10	.82405
6.63496	3.841	.05	.73105
6.63496	5.412	.02	.59850
6.63496	6.635	.01	.50000
6.63496	10.827	.001	.23743

Note: Critical χ^2 values for *df* = 1 from Table A.4.

MacCallum, Browne, and Sugawara (1996) provided a different approach to testing model fit using power and the root-mean-square error of approximation (RMSEA). Their approach also emphasized confidence intervals around RMSEA, rather than a single point estimate, so they suggested null and alternative values for RMSEA (exact fit: $H_o = .00$ versus $H_a = .05$; Close fit: $H_o = .05$ versus $H_a = .08$; and not close fit: $H_o = .05$ versus $H_a = .10$); researchers can also select their own. The MacCallum et al. (1996) method tests power, given exact fit (H_o; RMSEA = 0), close fit (H_o, RMSEA $\leq$ .05), or not close fit (H_o, RMSEA $\geq$ .05); and included SAS programs for calculating power given sample size or sample size given power using RMSEA. RMSEA is calculated as

$$\text{RMSEA} = \sqrt{\chi^2_{\text{Model}} - df_{\text{Model}} / [(N-1)df_{\text{Model}}]}$$

or

$$\text{RMSEA} = \sqrt{NCP / [(N-1)df_{\text{model}}]}$$

For example, given NCP = 6.63496; N = 301; and df = 7, RMSEA = 0.056209:

$$RMSEA = \sqrt{6.63496 / [(300)7]}$$

$$RMSEA = \sqrt{6.63496 / 2100}$$

$$RMSEA = .056209$$

SAS syntax—RMSEA and power

```
data chapter7;
do obs = 1;
n = 301;
df = 7;
alpha = .05;
* change rmseaHo and rmseaHa values to correspond to exact,
close, and not close values;
rmseaHo = .05;
rmseaHa = .08;
ncpHo = (n-1)*df*rmseaHo*rmseaHo;
ncpHa = (n-1)*df*rmseaHa*rmseaHa;
chicrit = quantile('chisquare',1-alpha,df);
if rmseaHo < rmseaHa then power = 1 - PROBCHI(chicrit,df,ncpHa);
if rmseaHo > rmseaHa then power = PROBCHI(chicrit,df,ncpHa);
output;
end;
Proc print;
Var n df alpha rmseaHo rmseaHa ncpHo ncpHa chicrit power;
Run;
```

SPSS syntax—RMSEA and power

```
DATA LIST FREE / obs.
BEGIN DATA.
1
END DATA.
compute n = 301.
compute df = 7.
compute alpha = .05.
comment change rmseaHo and rmseaHa values to correspond with
exact, close, not close values.
compute rmseaHo = .05.
compute rmseaHa = .08.
compute ncpHo = (n-1)*df*rmseaHo*rmseaHo.
compute ncpHa = (n-1)*df*rmseaHa*rmseaHa.
```

```
compute chicrit = IDF.CHISQ(1-alpha,df).
do if (rmseaHo < rmseaHa).
compute power = 1 - NCDF.CHISQ(chicrit, df, ncpHa).
else if (rmseaHo > rmseaHa).
compute power = NCDF.CHISQ(chicrit, df, ncpHa).
end if.
formats chicrit ncpHo ncpHa power (f8.5).
List.
```

The resulting SAS or SPSS output for close fit was given as:

obs	n	df	alpha	rmseaH_o	rmseaH_a	ncpH_o	ncpH_a	chicrit	power
1.00	301	7	.05	.05	.08	5.25	13.44	14.0671	.76813

We ran the recommended RMSEA values given by MacCallum et al. (1996) and listed them in Table 7.3. For exact fit, power = .33; for close fit, power = .768; and for not close fit, power = .06 ~ .057. A RMSEA model-fit value between .05 and .08 can be considered an acceptable model-fit index, when reported with other fit indices, and given power = .768. Power is interpreted as the probability of retaining the null hypothesis, thus a close fit between sample covariance matrix and model-implied covariance matrix.

Table 7.3:   Data from MacCallum et al. (1996) Null and Alternative Values for RMSEA Test of Fit

MacCallum Test	Ho	Ha	Power
Exact	.00	.05	.33034
Close	.05	.08	.76813
Not Close	.05	.01	.05756

Model values: α = .05, df = 7, N = 301.

MacCallum, Lee, and Browne (2010) further discussed power analysis issues in SEM, while Hancock and French (2013) more recently presented an in-depth discussion of the null and alternative model testing process in SEM, especially in regard to the non-centrality parameter (NCP; λ) and root-mean-square error of approximation (RMSEA; ε). They provided sample size tables for power and values of RMSEA, which can be produced using the Preacher and Coffman (2006) online calculator that provides the R code. Recall that the chi-square distribution, sample size, model degrees of freedom, and minimum fit function are all related in their formula expressions to yield RMSEA:

$$\chi^2 = (n-1)F_{ML}$$
$$\lambda = (n-1)F_{ML} \text{ or } \lambda = \chi^2 - df$$

and

$$\varepsilon = \sqrt{\frac{\chi^2 - df}{df(n-1)}}$$

The Preacher and Coffman (2006) online calculator generates R code for the calculations, which can be submitted directly online, copied into R directly, and/ or saved for future use. For example, to compute power = .80 (ability to retain RMSEA) given alpha = .05, $df = 8$, $N = 1200$, null RMSEA = .05, and alternative RMSEA = .02, the resulting calculation was power = .81. We can also compute sample size needed for a given power and RMSEA null and alternative values. For example, given alpha = .05, $df = 8$, power = .80, null RMSEA = .05, and alternative RMSEA = .02, the sample size = 1181. The R code for these examples is on the book website.

Sample Size

Sample size in SEM was previously indicated by Hoelter (1983) as the Critical N (CN) statistic, where $CN \geq 200$ was considered adequate. The Critical N is calculated as

$$CN = (\chi^2_{critical}/F_{min}) + 1$$

The critical chi-square ($\chi^2_{critical}$) is obtained for the model degrees of freedom at the .05 level of significance. The CN statistic is output by most SEM software packages. For example, given model $F_{min} = .0353432$ and $\chi^2_{critical} = 14.067$ for $df = 7$ at .05 level of significance (see Table 7.4), $CN = (14.067/.0353432) + 1 = 399$. CN gives the sample size at which the F_{min} value leads to a rejection of H_o. For a further discussion about CN refer to Bollen and Liang (1988) or Bollen (1989).

Sample size influences the calculation of the minimum fit function χ^2. Recall that the Minimum Fit Function χ^2 is computed as

```
Minimum Fit Function χ² = (N - g) × Fₘᵢₙ
```

Sample size also influences the calculation of the F_{min} values as follows:

```
Fₘᵢₙ = Minimum Fit Function χ²/(N - g)
```

The F_{min} is computed using the minimum fit function χ^2 in the computer output, sample size (N), and number of groups (g); while the non-centrality parameter (NCP) is computed using the Normal Theory χ^2 minus the degrees of freedom in the model. NCP is therefore computed as

```
NCP = Normal Theory Weighted Least Squares χ² - dfₘₒdₑₗ
```

The estimated sample size (N) using the previous NCP, $F_{min} = .046420$, $g = 1$ (single group), and the sample size of $N = 144$ is

$$N = (NCP/F_{min}) + g$$
$$= (6.63496/.046420) + 1$$
$$= 143.93 \sim 144$$

The F_{min} is calculated using the Minimum Fit Function χ^2, but NCP is calculated using the Normal Theory Weighted Least Squares χ^2. Be careful about these measures because SEM software calculates some measures of fit (NCP, RMSEA, and Independence model χ^2) using the normal theory weighted least squares χ^2, but uses the minimum fit function χ^2 for others. Differences between these two can be small if the multivariate normality assumption holds, or very large if not. For example, LISREL outputs four different χ^2 values: C1 = minimum fit function χ^2; C2 = normal theory weighted least squares χ^2; C3 = Satorra–Bentler scaled χ^2; and C4 = χ^2 corrected for non-normality (see Chapter Footnote).

To determine sample size for given df, alpha, and power for a theoretical model, the F_{min} value would be a fixed value from your model, but the NCP value would vary, as noted before. For example, the SAS program can be run for different NCP values to obtain corresponding sample size and power estimates. (*Note*: We are changing values of power in the SAS syntax program, but you can also fix power and change alpha values to obtain different sample sizes for different alpha levels at a specified power level, for example, power = .80.)

SAS syntax—sample size

```
data chapter7;
do obs = 1;
g = 1;
* change values of alpha to obtain sample size for given power;
alpha = .05;
fmin = .046420;
df = 1;
* change values of power to obtain sample size for given alpha;
power = .60;
chicrit = quantile('chisquare',1 - alpha, df);
ncp = CINV(power,df,chicrit);
n = (ncp/fmin) + g;
output;
end;
proc print;
var power n alpha ncp fmin g;
run;
```

The output from this first run with power = .60 would look like this:

obs	power	n	alpha	ncp	fmin	g	chicrit
1	.6	106.535	.05	4.89892	.04642	1	3.84146

We created Table 7.4 by changing the value of power for alpha = .05 for a critical χ^2 = 3.841, df = 1. (*Note: fmin* is fixed at .04642; *alpha* is fixed at .05, so *chicrit* will be fixed at 3.84146.) A sample size of N = 144 for power = .73 was correctly computed and indicated in the table. We see in Table 7.4 that sample size requirements increase as power increases, which is expected. Recall that NCP = $\chi^2 - df_{model}$, so the non-centrality parameter (NCP) is affected by the model chi-square and the degrees of freedom, which indicates a certain level of model complexity.

Table 7.4: Sample Size for Given Power with Alpha = .05

Power	n	Alpha	ncp	fmin	g	$\chi^2_{critical}$
.60	106.535	.05	4.89892	.04642	1	3.84146
.70	133.963	.05	6.17213	.04642	1	3.84146
.73	143.594	.05	6.61923	.04642	1	3.84146
.80	170.084	.05	7.84890	.04642	1	3.84146
.90	227.356	.05	10.5074	.04642	1	3.84146
.95	280.938	.05	12.9947	.04642	1	3.84146

Note: n should be rounded up, for example, 106.535 ~ 107.

We can also use the **SAS syntax—sample size** program to examine how changing the level of significance affects sample size for a fixed power value. Recall that F_{min} is fixed at .04642. Table 7.5 contains the output from the SAS program. We see in Table 7.5 that sample size requirements increase as the level of significance (alpha) for testing our model decreases, which is expected.

We used G*Power 3 to calculate various NCP values given alpha and power because SPSS did not have a command function to determine the non-centrality parameter (NCP) given power, *df*, and critical χ^2. (*Note*: SAS, R, STATA and other statistical programs have this capability.) In the *Test family* drop-down menu, select "χ^2 *test*"; in the *Statistical Test* drop-down menu select, "*Generic χ^2 test*"; and in the *Type of power analysis*, select "*Sensitivity: Compute non-centrality parameter – given α, and power.*" In the *Input Parameters* boxes, change the power value to .80 and the *df* value to 1. Click on the *Calculate* button; the *Output Parameters*, "*Critical χ^2*" and "*Non-centrality parameter γ*" will appear. The G*Power 3 dialog box should now display the Critical χ^2 = 3.84146 (associated with alpha = .05, *df* =1) and corresponding non-centrality parameter for power = .80. Table 7.5 reports these same values using the program **SAS syntax—sample size**.

Table 7.5: Sample Size for Given Alpha with Power = .80

Power	n	Alpha	ncp	fmin	g	χ^2_{critical}
.8	134.194	.10	6.18288	.04642	1	2.70554
.8	170.084	.05	7.84890	.04642	1	3.84146
.8	217.201	.02	10.0360	.04642	1	5.41189
.8	252.593	.01	11.6790	.04642	1	6.63490
.8	368.830	.001	17.0746	.04642	1	10.8276

Note: χ^2_{critical} values correspond to alpha values in Table A.4.

In planning a study, we should determine a priori what our sample size and power values should be. After gathering our data and running our SEM model (and any modifications), we should compute the post-hoc power using our non-centrality parameter from the computer output or sample size (N) using NCP and model F_{min} values. This should be easy given that $N = (\text{NCP}/F_{\text{min}}) + g$. We can *a-priori* specify values or obtain the F_{min} value from our model, calculate NCP using SAS or G*power 3 for a given *df*, critical chi-square, power, then use these values to calculate sample size (N). In the factor analysis chapter, we learned that 20 subjects per observed variables provided a good estimate, which is also reported in several multivariate statistics books, but now we have tools to help determine sample size and power in SEM.

Model Comparison

Model comparison involves comparing fit between alternative models, or testing parameter coefficients for significance between two alternative models. These hypothesis testing methods should involve constrained models with fewer parameters than the initial model. The initial (full) model represents the null hypothesis (H_o) and the alternative (constrained) model with fewer parameters is denoted H_a. Each model generates a χ^2 goodness-of-fit measure, and the difference between the models for significance testing is computed as $D^2 = \chi^2_o - \chi^2_a$, with $df_d = df_o - df_a$. The D^2 statistic is tested for significance at a specified alpha level (probability of Type I error), where H_o is rejected if D^2 exceeds the critical tabled χ^2 value with df_d degrees of freedom. The chi-square difference test or likelihood ratio test is used with GLS, ML, and WLS estimation methods.

A likelihood ratio (LR) test is possible between alternative models to examine the difference in χ^2 values between the initial model and a modified model. The LR test with degrees of freedom equal to $df_{\text{Initial}} - df_{\text{Modified}}$ is calculated as:

$$\text{LR} = \chi^2_{\text{Initial}} - \chi^2_{\text{Modified}}$$

For example, if Model A had $\chi^2 = 24.28099$, $df = 8$, and Model B had $\chi^2 = 13.92604$, $df = 7$; LR = 10.35495 with $df = 1$, which is a statistically significant chi-square value at the .05 level of significance ($\chi^2 = 3.84$, $df = 1$, $\alpha = .05$), indicating that the models are different. In this chi-square test, we may be seeking a significant chi-square value to indicate that our models are different.

$$LR_{df=1} = 24.28099 - 13.92604 = 10.35495$$

The LR test between models is possible when adding or dropping a single parameter (path or variable). A researcher will most likely be guided by modification indices with their associated change (decrease) in chi-square when modifying a model. For example, in the last chapter, the SIMPLIS modification indices were used to *add* an error covariance between *lozenges* and *cubes*.

Effect Size

MacCallum, Browne, and Cai (2006) presented an approach to compare nested models when the between model degrees of freedom are ≥ 1. They showed that when testing close fit, power results may differ depending upon the degrees of freedom in each model. Basically, the power to detect differences will be greater when models being compared have more degrees of freedom. For any given sample size, power increases as the model degrees of freedom increases. They defined an effect size (δ) in terms of model RMSEA and degrees of freedom for the two models.

Model A: $\varepsilon_A \sqrt{\dfrac{F_A}{df_A}}$

Model B: $\varepsilon_B \sqrt{\dfrac{F_B}{df_B}}$

Effect size index: $\delta = (F_A - F_B) = (df_A \varepsilon_A^2 - df_B \varepsilon_B^2)$

Non-centrality parameter: $\lambda = (N-1)(F_A - F_B) = (N-1)(df_A \varepsilon_A^2 - df_B \varepsilon_B^2)$

For our example values, the effect size (δ) would be computed as

$\delta = (df_{\text{Initial}} * \text{RMSEA}^2_{\text{Initial}} - df_{\text{Modified}} * \text{RMSEA}^2_{\text{Modified}})$
$\delta = ([8 * (.080937)^2] - [7 * (.046420)^2])$

$$\delta = (.0524056 - .0150836)$$
$$\delta = .037322$$

The non-centrality parameter is computed as

$$NCP = (N - 1)\,\delta$$

So, for our example:

$$NCP = (301 - 1) * (.037322)$$
$$NCP = 11.1966$$

Using G*Power 3, we enter this NCP = 11.1966, .05 level of significance, and df = 1 (model degree of freedom difference) and obtained power = .917. Power to detect a difference in RMSEA values is therefore possible for a given sample size with various degrees of freedom. The SAS program below also provides an ability to make power comparisons for different model degrees of freedom using RMSEA values from two nested models.

SAS syntax—effect size, RMSEA, and power

```
data chapter7;
 do obs = 1;
 n = 301;
 alpha = .05;
 dfa = 8;
 rmseaA = .080937;
 dfb = 7;
 rmseaB = .046420;
 delta = (dfa*rmseaA*rmseaA) - (dfb*rmseaB*rmseaB);
 ncp = (n - 1)*delta;
 dfdiff = dfa - dfb;
 chicrit = quantile('chisquare',1 - alpha, dfdiff);
 power = 1 - PROBCHI(chicrit, dfdiff,ncp);
 output;
 end;
 Proc print;
 var n dfa rmseaA dfb rmseaB delta ncp dfdiff chicrit power;
 run;
```

The computer output should look like this:

Obs	n	dfa	rmseaA	dfb	rmseaB	delta	ncp	dfdiff	chicrit	power
1	301	8	0.080937	7	0.04642	.037323	11.1968	1	3.84146	.91716

The power = .91716 indicates a 91% chance of detecting a difference between the model RMSEA values.

Power is affected by the size of the model degrees of freedom (degrees of freedom implies a certain degree of model complexity). The G*Power 3 program or the SAS program can be used for models where the difference in degrees of freedom is greater than one. We therefore ran a comparison for a model with different levels of degrees of freedom to show how power is affected. In Table 7.6, power increases dramatically when the level of degrees of freedom increases from 5 to 14 while maintaining a model degrees of freedom difference at $df = 1$. You can also output program values for $df \geq 2$ to see the effect on power.

Table 7.6: Data from MacCallum et al. (2006): Power at Increasing Model Degrees of Freedom

dfa	dfb	Power
5	4	.76756
8	7	.91716*
11	10	.97337
14	13	.99206

(RMSEA approach)
Model values: $\alpha = .05$, N = 301.

PARAMETER SIGNIFICANCE

A single parameter can be tested for significance using nested models. *Nested models* involve an initial model being compared to a modified model in which a single parameter has been fixed to zero (dropped) or estimated (added). In structural equation modeling, the intent is to determine the significance of the decrease in the χ^2 value for the modified model from the initial model. The LR test is used to test the difference in the models for a single added parameter, for example, adding error covariance between *lozenges* and *cubes* in the previous chapter on factor analysis.

Power can be computed for testing the significance of an individual parameter estimate. For GLS, ML, and WLS estimation methods, this involves determining the significance of χ^2 with one degree of freedom ($\chi^2 = 3.84$, $df = 1$, $\alpha = .05$) for a single parameter estimate, thus determining the significance of the reduction in χ^2 that should equal or exceed the modification index value for the parameter estimate fixed to zero. Power values for modification index values can be computed using SAS because the modification index (MI) is a non-centrality parameter (NCP). The power of an MI value (NCP) at the .05 level of significance, $df = 1$, critical chi-square value = 3.841 is computed in the following SAS syntax program for MI = 10.4. Power = .89, so in testing the statistical significance of MI for the added parameter (error covariance), we have an 89% chance of correctly rejecting the null hypothesis and accepting the alternative hypothesis that MI is different from zero. We desire MI to be statistically significant, thus indicating it should be added to the model.

SAS syntax—power for parameter MI value

```
data chapter7;
 do obs = 1;
 mi = 10.4;
 alpha = .05;
 df = 1;
 chicrit = quantile('chisquare',1 - alpha, df);
 power = 1 - PROBCHI(chicrit, df, mi);
 output;
 end;
 Proc print;
 var mi power alpha df chicrit;
 run;
```

The SAS output indicated the following:

Obs	mi	power	alpha	df	chicrit
1	10.4	.89075	.05	1	3.84146

Power values for parameter estimates can also be computed using an SAS program because a squared t or z value for a parameter estimate is asymptotically distributed as a non-central chi-square, that is, NCP = T^2. So, if a model indicated an error covariance = 8.34, with standard error = 2.62, then t = 8.34/2.62 = 3.19. The power of a squared T value for the parameter estimate is computed in an SAS program as follows:

SAS syntax—power for parameter T value

```
data chapter7;
 do obs = 1;
 T = 3.19;
 ncp = T*T;
 alpha = .05;
 df = 1;
 chicrit = quantile('chisquare',1 - alpha, df);
```

```
power = 1 - PROBCHI(chicrit, df, ncp);
output;
end;
Proc print;
var ncp power alpha df chicrit;
run;
```

The SAS output looks like this:

Obs	ncp	power	alpha	df	chicrit
1	10.1761	.89066	.05	1	3.84146

Power = .89, so in testing the statistical significance of the parameter estimate, we have an 89% chance of correctly rejecting the null hypothesis and accepting the alternative hypothesis that T is different from zero. Here again, a traditional power interpretation is given where we want to reject the null hypothesis and accept the alternative hypothesis, which indicates that NCP (T^2) is statistically significant. The other model-fit indices (GFI, AGFI, NFI, IFI, CFI, etc.) do not have a test of statistical significance and therefore do not involve power calculations.

In SEM, we can use SAS, SPSS, Gpower, online calculators, or R to plan our model analysis using sample size, power, alpha level, effect size, null and alternative RMSEA values, or NCP (non-centrality). As in traditional statistics, a researcher now has tools to guide whether they have a sufficient sample size and/or power to detect a model fit or an alternative model difference. A better understanding of the role chi-square, degrees of freedom, and F minimum fit function have in calculating the non-centrality parameter (NCP) and root-mean-square error of approximation (RMSEA) also helps in understanding sample size or power requirements in SEM applications. A researcher can now *a-priori* (plan ahead) or *a-posteriori* (afterwards) obtain sample size and power.

Summary

Research suggests that certain model-fit indices are more susceptible to sample size than others, hence, power. We have already learned that χ^2 is affected by sample size, that is, $\chi^2 = (N - 1) F_{ML}$, where F_{ML} is the maximum likelihood fit function for a model, and therefore χ^2 increases in direct relation to $N - 1$ (Bollen, 1989). Kaplan (1995) also pointed out that power in SEM is affected by the size of the misspecified parameter, sample size, and location of the parameter in the model. Specification errors induce bias in the standard errors and parameter estimates, and thus affect power. These factors also affect power in other parametric statistical tests (Cohen, 1988). Saris and Satorra (1993) pointed out that the larger the non-centrality parameter, the greater is the power of the test, that is, an evaluation of the power of the test is an evaluation of the non-centrality parameter.

Muthén and Muthén (2002) outlined how Monte Carlo methods can be used to decide on the power for a given specified model using the Mplus program. Power is indicated as the percentage of significant coefficients or the proportion of replications for which the null hypothesis that a parameter is equal to zero is rejected at the .05 level of significance, two-tailed test, with a critical value of 1.96. The authors suggested that power values equal or exceed the traditional .80 level for determining the probability of rejecting the null hypothesis when it is false.

Marsh et al. (1988, 1996) also examined the influence of sample size on 30 different model-fit indices and found that the Tucker–Lewis index (Tucker & Lewis, 1973) and four new indices based on the Tucker–Lewis index were the only ones relatively independent of sample size. Bollen (1990) argued that the claims regarding which model-fit indices were affected by sample size needed further clarification. There are actually two sample size effects that are confounded: (a) whether sample size enters into the calculation of the model-fit index, and (b) whether the means of the sampling distribution of the model-fit index are related to sample size. Sample size was shown not to affect the calculation of NFI, TLI, GFI, AGFI, and CN, but the means of the sampling distribution of these model-fit indices were related to sample size. Bollen (1990) concluded that, given a lack of consensus on the best measure of fit, it is prudent to report multiple measures rather than to rely on a single choice; we concur.

Muthén and Muthén (2002) also used Mplus to determine appropriate sample sizes in the presence of model complexity, distribution of variables, missing data, reliability, and variance–covariance of variables. For example, given a two-factor CFA model and 10 indicator variables with normally distributed non-missing data, a sample size of 150 is indicated with power = .81. In the presence of missing data, sample size increases to $n = 175$. Given non-normal missing data, sample size increases to $n = 315$. Davey and Savla (2009) provide an excellent treatment of statistical power analysis with missing data via a structural equation modeling approach. Their examples cover many different types of modeling situations using SAS, STATA, SPSS, or LISREL syntax programs. This is a must-read book on the subject of power and sample size, especially in the presence of missing data.

Finally, one should beware of claims of sample size effects on fit measures that do not distinguish the type of sample size effect (Satorra & Bentler, 1994). Cudeck and Henly (1991) also argued that a uniformly negative view of the effects of sample size in model selection is unwarranted. They focused instead on the predictive validity of models in the sense of cross-validation in future samples while acknowledging that sample size issues are a problem in the field of statistics in general and unavoidable in structural equation modeling.

TWO-STEP VERSUS FOUR-STEP APPROACH TO MODELING

Anderson and Gerbing (1988) proposed a two-step model-building approach that emphasized the analysis of two conceptually distinct models: a measurement model followed by the structural model (Lomax, 1982). The *measurement* model, or factor model, specifies the relationships among measured (observed) variables underlying the latent variables. The *structural* model specifies relationships among the latent variables as posited by theory. The measurement model provides an assessment of convergent and discriminant validity, and the structural model provides an assessment of nomological validity.

Mulaik et al. (1989) expanded the idea of model fit by assessing the relative fit of the structural model among latent variables, independently of assessing the fit of the indicator variables in the measurement model. The relative normed fit index (RNFI) makes the following adjustment to separately estimate the effects of the structural model from the measurement model: $\text{RNFI}_j = (F_u - F_j)/[F_u - F_m - (df_j - df_m)]$, where $F_u = \chi^2$ of the full model, $F_j = \chi^2$ of the structural model, $F_m = \chi^2$ of the measurement model, df_j is the degrees of freedom for the structural model, and df_m is the degrees of freedom for the measurement model. A corresponding relative parsimony ratio (RP) is given by $RP_j = (df_j - df_m)/(df_u - df_m)$, where df_j is the degrees of freedom for the structural model, df_m is the degrees of freedom for the measurement model, and df_u is the degrees of freedom for the null model. In comparing different models for fit, Mulaik et al. multiplied RP_j by RNFI_j to obtain a relative parsimony fit index appropriate for assessing how well and to what degree the models explained both relationships in the measurement of latent variables and the structural relationships among the latent variables by themselves. McDonald and Marsh (1990), however, doubted whether model parsimony and goodness of fit could be captured by this multiplicative form because it is not a monotonic increasing function of model complexity. Obviously, further research will be needed to clarify these issues.

Mulaik and Millsap (2000) presented a four-step approach to testing a nested sequence of SEM models:

Step 1: Specify an unrestricted measurement model, namely conducting an exploratory common factor analysis to determine the number of factors (latent variables) that fit the variance–covariance matrix of the observed variables.

Step 2: Specify a confirmatory factor model that tests hypotheses about certain relations among indicator variables and latent variables. Basically, certain factor loadings are fixed to zero in an attempt to have only a single non-zero factor loading for each indicator variable of a latent variable. Sometimes this leads to a lack of measurement model fit because an indicator variable may have a relation with another latent variable.

Step 3: Specify relations among the latent variables in a structural model. Certain relations among the latent variables are fixed to zero so that some latent variables are not related to one another.

Step 4: Determine the acceptable fit of the structural model, that is, CFI > .95 and RMSEA < .05. A researcher plans hypotheses about free parameters in the model. Several approaches are possible: (a) perform simultaneous tests in which free parameters are fixed based on theory or estimates obtained from other research studies; (b) impose fixed parameter values on freed parameters in a nested sequence of models until a misspecified model is achieved (misspecified parameter); or (c) perform a sequence of confidence-interval tests around free parameters using the standard errors of the estimated parameters.

We agree with the basic Mulaik and Millsap (2000) approach and recommend that the measurement models for latent variables be established first and then structural models establishing relations among the latent independent and dependent variables be formed. It is in the formulation of measurement models that most of the model modifications occur to obtain acceptable data to model fit. In fact, a researcher could begin model generation by using exploratory factor analysis (EFA) on a sample of data to find the number and type of latent variables in a plausible model (Costello & Osborne, 2005). Once a plausible model is determined, another sample of data could be used to confirm or test the factor model, that is, confirmatory factor analysis (CFA) (Jöreskog, 1969). Exploratory factor analysis is even recommended as a precursor to confirmatory factor analysis when the researcher does not have a substantive theoretical model (Gerbing & Hamilton, 1996).

Measurement invariance is also important to examine, which refers to considering similar measurement models across different groups; for example, does the factor (latent variable) imply the same thing to men and women? This usually involves adding between group constraints in the measurement model. If measurement invariance cannot be established, then the finding of a between-group difference is questionable (Cheung & Rensvold, 2002). Cheung and Rensvold (2002) also recommend that the comparative fit index (CFI), gamma hat, and

McDonald's non-centrality index (NCI) be used for testing between group measurement invariance of CFA models rather than the goodness-of-fit index (GFI) or the likelihood ratio test (LR), also known as the chi-square difference test. Byrne and Watkins (2003) questioned whether measurement invariance could be established given that individual items on an instrument could exhibit invariance or group differences. Byrne and Sunita (2006) provided a step-by-step approach for examining measurement invariance.

SUMMARY

A researcher generally conducts SEM using three basic approaches: confirmatory models, alternative models, and model generation. A determination of their model fit generally falls into one of three categories: model fit, model comparison, and model parsimony. AMOS has provided an innovative approach to finding a model that fits the data, called specification search.

We examined in detail the different categories of model-fit criteria because different fit indices have been developed depending on the type of model tested. Generally, no single model-fit index is sufficient for testing a hypothesized structural model. An ideal fit index just does not exist. This is not surprising because it has been suggested that an ideal fit index is one that is independent of sample size, accurately reflects differences in fit, imposes a penalty for inclusion of additional parameters (Marsh et al., 1988), and supports the choice of a true model when it is known (McDonald & Marsh, 1990). Researchers generally are guided by modification indices when making a change to a poor fitting model. Specification search techniques, Tabu and optimization algorithms, as well as the specification search in AMOS (SPSS, 2009), makes finding a model fit easier.

The SEM modeling steps (specification, identification, estimation, testing, and modification) will become clearer in the following chapters with SEM modeling applications. This chapter explained the *basics* of structural equation modeling. In subsequent chapters, we will present various SEM models that demonstrate the variety of applications suitable for structural equation modeling. You are encouraged to explore other examples and applications reported in SEM books (Marcoulides & Schumacker, 1996, 2001), the SEM software user guides, and the references at the end of this chapter. Our intention is to provide a basic understanding of the SEM applications to further your interest in the structural equation modeling approach.

EXERCISES

1. Define the following SEM modeling steps:
   a. Model specification
   b. Model identification
   c. Model estimation
   d. Model testing
   e. Model modification.
2. Define confirmatory models, alternative models, and model-generating approaches.
3. Define model fit, model comparison, and model parsimony.
4. Calculate the following fit indices for the model output in Figure 6.2 (Chapter 6):

   GFI $= 1 - (\chi^2_{model}/\chi^2_{null})$

   NFI $= (\chi^2_{null} - \chi^2_{model})/\chi^2_{null}$

   RFI $= 1 - [(\chi^2_{model}/df_{model})/(\chi^2_{null}/df_{null})]$

   IFI $= (\chi^2_{null} - \chi^2_{model})/(\chi^2_{null} - df_{model})$

   TLI $= [(\chi^2_{null}/df_{null}) - (\chi^2_{model}/df_{model})]/[(\chi^2_{null}/df_{null}) - 1]$

   CFI $= 1 - [(\chi^2_{model} - df_{model})/(\chi^2_{null} - df_{null})]$

   Model AIC $= \chi^2_{model} + 2q$ (Note: $q = df - 1$)

   Null AIC $= \chi^2_{null} + 2q$ (Note: $q = df - 1$)

   $$RMSEA = \sqrt{[\chi^2_{Model} - df_{Model}]/[(N-1)df_{Model}]}$$

5. How are modification indices in LISREL-SIMPLIS used?
6. What steps should a researcher take in examining parameter estimates in a model?
7. How should a researcher test for the difference between two alternative models?
8. How are structural equation models affected by sample size and power considerations?
9. Describe the SEM modeling steps.
10. What new approaches are available to help a researcher identify the best model?
11. Use G*Power 3 to calculate power for a model with NCP = 6.3496 at $p = .05$, $p = .01$, and $p = .001$ levels of significance. What happens to power when alpha increases?
12. Use G*Power 3 to calculate power for a modified model with alpha = .05 and NCP = 6.3496 at $df = 1$, $df = 2$, and $df = 3$ levels of model complexity. What happens to power when degrees of freedom increases?

CHAPTER FOOTNOTE

SEM software packages may output different chi-square values. For example, LISREL computes two different sets of standard errors for parameter estimates and up to four different chi-squares for testing overall fit of the model. These new standard errors and chi-squares can be obtained for single-group problems as well as multiple-group problems using variance–covariance matrices with or without means. Which standard errors and which chi-squares will be reported depends on whether an asymptotic covariance matrix is provided and which method of estimation is used to fit the model (ULS, GLS, ML, WLS, DWLS). The asymptotic covariance matrix is a consistent estimate of N times the asymptotic covariance matrix of the sample matrix being analyzed. Standard errors are estimated under non-normality if an asymptotic covariance matrix is used. Standard errors are estimated under multivariate normality if no asymptotic covariance matrix is used.

Chi-squares

Four different chi-squares are reported in LISREL and denoted below as C1, C2, C3, and C4, where the × indicates that it is reported for any of the five estimation methods.

Asymptotic covariance matrix not provided:

	ULS	GLS	ML	WLS	DWLS
C1	—	×	×	—	—
C2	×	×	×	—	—
C3	—	—	—	—	—
C4	—	—	—	—	—

Asymptotic covariance matrix provided:

	ULS	GLS	ML	WLS	DWLS
C1	—	×	×	×	—
C2	×	×	×	—	×
C3	×	×	×	—	×
C4	×	×	×	—	×

Note 1: C1 is $n - 1$ times the minimum value of the fit function; C2 is $n - 1$ times the minimum of the WLS fit function using a weight matrix estimated under multivariate normality; C3 is the Satorra–Bentler scaled chi-square statistic or its generalization to mean and covariance structures and multiple groups (Satorra & Bentler, 1994); C4 is computed from equations in Browne (1984) or Satorra (1993) using the asymptotic covariance matrix.

The corresponding chi-squares are now given in the output as follows:

C1: Minimum fit function chi-square
C2: Normal theory weighted least squares chi-square
C3: Satorra–Bentler scaled chi-square
C4: Chi-square corrected for non-normality

Note 2: Under multivariate normality of the observed variables, C1 and C2 are asymptotically equivalent and have an asymptotic chi-square distribution if the model holds exactly and an asymptotic noncentral chi-square distribution if the model holds approximately. Under normality and non-normality, C2 and C4 are correct asymptotic chi-squares, but may not be the best chi-square in small and moderate samples. Hu, Bentler, and Kano (1992) and Yuan and Bentler (1997) found that C3 performed better given different types of models, sample size, and degrees of non-normality.

REFERENCES

Akaike, H. (1987). Factor analysis and AIC. *Psychometrika*, 52, 317–332.

Anderson, J. C., & Gerbing, D. W. (1984). The effects of sampling error on convergence, improper solutions and goodness-of-fit indices for maximum likelihood confirmatory factor analysis. *Psychometrika*, 49, 155–173.

Anderson, J. C., & Gerbing, D. W. (1988). Structural equation modeling in practice: A review and recommended two-step approach. *Psychological Bulletin*, 103, 411–423.

Baldwin, B. (1989). A primer in the use and interpretation of structural equation models. *Measurement and Evaluation in Counseling and Development*, 22, 100–112.

Bentler, P. M. (1990). Comparative fit indexes in structural models. *Psychological Bulletin*, 107, 238–246.

Bentler, P. M., & Bonett, D. G. (1980). Significance tests and goodness-of-fit in the analysis of covariance structures. *Psychological Bulletin*, 88, 588–606.

Bollen, K. A. (1989). *Structural equations with latent variables*. New York: Wiley.

Bollen, K. A. (1990). Overall fit in covariance structure models: Two types of sample size effects. *Psychological Bulletin*, 107, 256–259.

Bollen, K. A., & Liang, J. (1988). Some properties of Hoelter's CN. *Sociological Methods and Research*, 16, 492–503.

Bollen, K. A., & Long, S. J. (1993). *Testing structural equation models*. Newbury Park, CA: Sage.

Browne, M. W. (1984). Asymptotically distribution-free methods for the analysis of covariance structures. *British Journal of Mathematical and Statistical Psychology*, 37, 62–83.

Browne, M. W., & Cudeck, R. (1993). Alternative ways of assessing model fit. In K. A. Bollen, & J. S. Long (Eds.), *Testing structural equation models* (pp. 132–162). Beverly Hills, CA: Sage.

Byrne, B., & Sunita, M. S. (2006). The MACS approach to testing for multigroup invariance of a second-order structure—A walk through the process. *Structural Equation Modeling: A Multidisciplinary Journal,* 13(2), 287–321.

Byrne, B. M., & Watkins, D. (2003). The issue of measurement invariance revisited. *Journal of Cross-Cultural Psychology*, 34(2), 155–175.

Cheung, G. W., & Rensvold, R. B. (2002). Evaluating goodness-of-fit indexes for testing measurement invariance. *Structural Equation Modeling*, 9, 233–255.

Cliff, N. (1983). Some cautions concerning the application of causal modeling methods. *Multivariate Behavioral Research*, 18, 115–126.

Cohen, J. (1988). *Statistical power analysis for the behavioral sciences* (2nd edn.). Hillsdale, NJ: Lawrence Erlbaum.

Costello, A. B., & Osborne, J. (2005). Best practices in exploratory factor analysis: Four recommendations for getting the most from your analysis. *Practical Assessment Research & Evaluation*, 10(7), 1–9.

Cudeck, R., & Henly, S. J. (1991). Model selection in covariance structure analysis and the "problem" of sample size: A clarification. *Psychological Bulletin*, 109, 512–519.

Davey, A., & Savla, J. (2009). *Statistical power analysis with missing data: A structural equation modeling approach.* New York: Routledge/Taylor & Francis.

Faul, F., Erdfelder, E., Lang, A.-G., & Buchner, A. (2007). G*Power 3: A flexible statistical power analysis program for the social, behavioral, and biomedical sciences. *Behavior Research Methods,* 39, 175–191.

Gerbing, D. W., & Hamilton, J. G. (1996). Viability of exploratory factor analysis as a precursor to confirmatory factor analysis, *Structural Equation Modeling,* 3(1), 62–72.

Gonzalez, R., & Griffin, D. (2001). Testing parameters in structural equation modeling: Every "one" matters. *Psychological Methods*, 6(3), 258–269.

Hair, J. F., Jr., Anderson, R. E., Tatham, R. L., & Black, W. C. (1992). *Multivariate data analysis with readings* (3rd edn.). New York: Macmillan.

Hancock, G. R. & French, B. (2013). *Power Analysis in Structural Equation Modeling.* In G. R. Hancock, & R. O. Mueller (Eds.), *Structural equation modeling: A second course* (2nd edn., pp. 117–159). Greenwich, CT: Information Age.

Hershberger, S. E. & Marcoulides, G. A. (2013). The problem of equivalent structural models. In G. R. Hancock, & R. O. Mueller (Eds.), *Structural equation modeling: A second course* (2nd edn., pp. 3–39). Greenwich, CT: Information Age.

Hoelter, J. W. (1983). The analysis of covariance structures: Goodness-of-fit indices. *Sociological Methods and Research*, 11, 325–344.

Hu, L., & Bentler, P. M. (1995). Evaluating model fit. In R. H. Hoyle (Ed.), *Structural equation modeling: Concepts, issues, and applications* (pp. 76–99). Thousand Oaks, CA: Sage.

Hu, L., Bentler, P. M., & Kano, Y. (1992). Can test statistics in covariance structure analysis be trusted? *Psychological Bulletin,* 112, 351–362.

James, L. R., Mulaik, S. A., & Brett, J. M. (1982). *Causal analysis: Assumptions, models, and data.* Beverly Hills, CA: Sage.

Jöreskog, K. G. (1969). A general approach to confirmatory maximum likelihood factor analysis. *Psychometrika,* 34, 183–202.

Jöreskog, K. G., & Sörbom, D. (1993). *LISREL 8: Structural equation modeling with the SIMPLIS command language.* Hillsdale, NJ: Lawrence Erlbaum.

Kaplan, D. (1995). Statistical power in structural equation modeling. In R. H. Hoyle (Ed.), *Structural equation modeling: Concepts, issues, and applications* (pp. 100–117). Thousand Oaks, CA: Sage.

Kenny, D. A., & McCoach, D. B. (2003). Effect of the number of variables on measures of fit in structural equation modeling. *Structural Equation Modeling,* 10, 333–351.

Loehlin, J. C. (1987). *Latent variable models: An introduction to factor, path, and structural analysis.* Hillsdale, NJ: Lawrence Erlbaum.

Lomax, R. G. (1982). A guide to LISREL-type structural equation modeling. *Behavior Research Methods and Instrumentation,* 14, 1–8.

Lunneborg, C. E. (1987). *Bootstrap applications for the behavioral sciences. Vol. 1.* Seattle: University of Washington, Psychology Department.

MacCallum, R. C., Browne, M. W., & Cai, L. (2006). Testing differences between nested covariance structure models: Power analysis and null hypotheses. *Psychological Methods,* 11, 19–35.

MacCallum, R. C., Browne, M. W., & Sugawara, H. M. (1996). Power analysis and determination of sample size for covariance structure modeling. *Psychological Methods,* 1, 130–149.

MacCallum, R. C., Lee, T., & Browne, M. W. (2010). The issue of isopower in power analysis for tests of structural equation models. *Structural Equation Modeling,* 17, 23–41.

Marcoulides, G. A., & Drezner, Z. (2001). Specification searches in structural equation modeling with a genetic algorithm. In G. A. Marcoulides & R. E. Schumacker (Eds.), *New developments and techniques in structural equation modeling* (pp. 247–268). Mahwah, NJ: Erlbaum.

Marcoulides, G. A., & Drezner, Z. (2003). Model specification searches using ant colony optimization algorithms. *Structural Equation Modeling,* 10, 154–164.

Marcoulides, G., & Schumacker, R. E. (Eds.). (1996). *Advanced structural equation modeling: Issues and techniques.* Mahwah, NJ: Lawrence Erlbaum Associates.

Marcoulides, G., & Schumacker, R. E. (Eds.). (2001). *New developments and techniques in structural equation modeling.* Mahwah, NJ: Lawrence Erlbaum Associates.

Marcoulides, G. A., Drezner, Z., & Schumacker, R. E. (1998). Model specification searches in structural equation modeling using Tabu search. *Structural Equation Modeling,* 5, 365–376.

Marsh, H. W., Balla, J. R., & Hau, K.-T. (1996). An evaluation of incremental fit indices: A clarification of mathematical and empirical properties. In G. A.

Marcoulides & R. E. Schumacker (Eds.), *Advanced structural equation modeling: Issues and techniques* (pp. 315–353). Mahwah, NJ: Lawrence Erlbaum.

Marsh, H. W., Balla, J. R., & McDonald, R. P. (1988). Goodness-of-fit indexes in confirmatory factor analysis: The effect of sample size. *Psychological Bulletin*, 103, 391–410.

McDonald, R. P. (1989). An index of goodness-of-fit based on noncentrality. *Journal of Classification*, 6, 97–103.

McDonald, R. P., & Marsh, H. W. (1990). Choosing a multivariate model: Noncentrality and goodness of fit. *Psychological Bulletin*, 107, 247–255.

Mulaik, S. A., & Millsap, R. E. (2000). Doing the four-step right. *Structural Equation Modeling*, 7, 36–73.

Mulaik, S. A., James, L. R., Alstine, J. V., Bennett, N., Lind, S., & Stilwell, C. D. (1989). Evaluation of goodness-of-fit indices for structural equation models. *Psychological Bulletin*, 105, 430–445.

Muthén, B., & Muthén, L. (2002). How to use a Monte Carlo study to decide on sample size and determine power. *Structural Equation Modeling*, 9, 599–620.

Preacher, K. J. & Coffman, D. I. (2006). Computing power and minimum sample size for RMSEA [Computer software]. Available at: quantpsy.org/calc.htm. Accessed July 2015.

Saris, W. E., & Satorra, A. (1993). Power evaluation in structural equation models. In K. Bollen, & J. S. Long (Eds.), *Testing structural equation models* (pp. 181–204). Newbury Park, CA: Sage.

Satorra, A. (1993). Multi-sample analysis of moment structures: Asymptotic validity of inferences based on second-order moments. In K. Haagen, D. J. Bartholomew, & M. Deistler (Eds.), *Statistical modeling and latent variables* (pp. 283–298). Amsterdam: Elsevier.

Satorra, A., & Bentler, P. M. (1994). Corrections for test statistics and standard errors in covariance structure analysis. In A. Von Eye, & C. C. Clogg (Eds.), *Latent variable analysis: Applications for developmental research* (pp. 399–419). Thousand Oaks, CA: Sage.

SPSS (2009). *Statistics 17.0*. SPSS, Inc.: Chicago, IL.

Steiger, J. H. (1990). Structural model evaluation and modification: An interval estimation approach. *Multivariate Behavioral Research*, 25, 173–180.

Steiger, J. H., & Lind, J. M. (1980, May). *Statistically-based tests for the number of common factors*. Paper presented at Psychometric Society Meeting, Iowa City, IA.

Tucker, L. R., & Lewis, C. (1973). The reliability coefficient for maximum likelihood factor analysis. *Psychometrika*, 38, 1–10.

Williams, L. J., & Holahan, P. J. (1994). Parsimony-based fit indices of multiple indicator models: Do they work? *Structural Equation Modeling: A Multidisciplinary Journal*, 1, 161–189.

Yuan, K.-H., & Bentler, P. M. (1997). Mean and covariance structure analysis: Theoretical and practical improvements. *Journal of the American Statistical Association*, 92, 767–774.

Chapter 8

MULTIPLE GROUP (SAMPLE) MODELS

CHAPTER CONCEPTS

Measurement model group difference
   Measurement invariance
Structural model group difference
Multiple sample model
Testing for parameter differences

Multiple group models were developed to test whether groups had the same or different theoretical model (Lomax, 1985). This approach can be applied in comparing groups on a measurement model, which is called a test of measurement invariance. *Measurement invariance* implies that two or more groups have the same measurement model, that is, the same construct or meaning for the latent variables in the measurement model. Byrne and Sunita (2006) provided a step-by-step approach for examining measurement invariance in SEM.

Once a researcher establishes that the latent variable constructs are the same, a comparison of the groups in the structural model can be conducted. A researcher would desire a non-significant chi-square difference between the groups in the measurement model invariance comparison. However, a researcher might want the structural models to be different, thus a significant chi-square difference. For example, adolescent drug use is different between high and low GPA high school students in the structural model, but the self concept and attitude toward drugs constructs in the measurement models are the same.

A researcher could also compare multiple samples of data on a measurement model, thus establishing the validity of a construct. You are basically applying a single specified model to one or more samples of data. Also, samples of data can be applied to a structural model to establish model validity. A unified approach to multi-group modeling is explained in Marcoulides and Schumacker (2001).

A misconception in multiple group modeling is that all paths in the model have to be the same. In fact, the chi-square statistic is a test of global fit; an overall fit of a model. It is possible for a measurement model to be slightly different for each group. Some parameters can therefore be different between the groups in the theoretical model. We can constrain or fix some parameters to be different between the groups, but allow other paths to be tested for equality. You might even consider dropping a single indicator variable that is different between the groups in a measurement model. This type of SEM modeling permits testing for group differences in the specified model without all parameters being tested for equality (Jöreskog & Sörbom, 1993). It is also possible to test for differences in specific parameter estimates.

MEASUREMENT MODEL GROUP DIFFERENCE

The global model fit is assessed using the chi-square test, which is dependent on sample size. The chi-square statistic will reject models if the sample size is large, that is, yield a significant chi-square value; and will fail to reject models if the sample size is too small, that is, yield a non-significant chi-square value. So, other types of fit indices were created to assess the fit of a model when making multiple group comparisons. The comparative fit index (CFI) or the Tucker–Lewis (TLI) index with values > .95 are considered acceptable. The root-mean-square error of approximation (RMSEA) with values less than .05 is considered acceptable, because it is insensitive to sample size, although affected by model complexity (larger degrees of freedom). The Akaike Information Criterion (AIC) and Bayesian Information Criterion (BIC) are both used to compare models using the –2LL chi-square fit value; however, they are sensitive to the number of parameters in a model (model complexity). The rule of thumb is to choose models with lower AIC or BIC values.

Our initial test of a theoretical measurement model is based on the Holzinger and Swineford (1939) study, where a confirmatory model of spatial and verbal ability was presented in Chapter 6. However, this time the raw data set (HS.data) will be used from the *MBESS* package in R with $N = 301$ observations and 32 variables [library(MBESS); data(HS.data)]. We need the raw data set to extract separate variance–covariance matrices on the demographic variable of interest, *school*, to be able to test for group differences in the measurement model. We have assessed the global model fit in Chapter 6 for the indicator variables of the constructs indicated in the theoretical measurement model diagrammed in Figure 8.1 ($\chi^2 = 13.68$, $df = 7$, $p = .06$). We concluded that the sample data fit the theoretical measurement model.

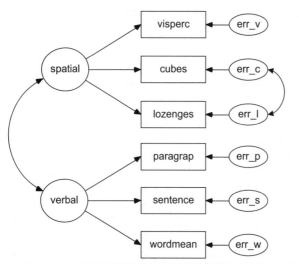

■ **Figure 8.1:** THEORETICAL MEASUREMENT MODEL (AMOS)

Measurement Invariance

The multiple group model analysis generally establishes that the data for each group fit the same measurement model. This is considered a test of *measurement invariance* for the groups, which means that the constructs are the same for each group. We would proceed by running separate models for each group to assess the data to model fit. It is possible that the two separate group models may not be exactly the same, that is, a few modification indices may be indicated for the error covariances in one group, but not in the other group. Differences in random error are acceptable, but we desire the factor loadings and factor correlations to be similar in the measurement model. The chi-square statistics for the two groups on the theoretical measurement model are compared. This results in an omnibus chi-square test of model fit for group differences, which we desire to be non-significant to indicate that the constructs are the same for both groups.

The individual SEM measurement model programs would contain different sample sizes and covariance matrices for each group, but have the same specified measurement model. The separate covariance matrices for each group are given in Table 8.1.

The results in Table 8.2 indicate that the Pasteur school and the Grant-White school fit the measurement model differently, as noted by the non-significant chi-square and the RMSEA, CFI, and GFI indices for each group. The

Table 8.1: Pasteur and Grant-White Covariance Matrices

Pasteur (n = 156)	VISPERC	CUBES	LOZENGES	PARCOMP	SENCOMP	WORDMEAN
	50.552316					
	9.712738	24.215219				
	29.957155	15.354673	86.686187			
	10.288627	1.068900	3.977337	11.953805		
	11.399628	2.295533	2.139950	13.041315	27.502854	
	21.032465	5.468073	12.517949	15.959843	26.318486	48.056038
Grant-White (n = 145)	VISPERC	CUBES	LOZENGES	PARCOMP	SENCOMP	WORDMEAN
	47.800958					
	10.012500	19.758333				
	25.797893	15.416667	69.172414			
	7.972605	3.420833	9.206657	11.393487		
	9.935728	3.295833	11.091954	11.277347	21.615709	
	17.425335	6.876389	22.954262	19.166523	25.320977	63.162548

Table 8.2: Comparison of Global and Individual School Measurement Models

Measurement Model	Global		Pasteur[1]		Grant-White	
	Spatial	Verbal	Spatial	Verbal	Spatial	Verbal
Factor loadings:						
Visual Perception	.96		.91		.68	
Cubes	.31		.33		.46	
Lozenges	.46		.50		.66	
Paragraph Comprehension		.85		.82		.87
Sentence Completion		.86		.86		.83
Word Meaning		.84		.83		.83
(Cubes, Lozenges)	.20			.25		.11
(Spatial, Verbal)	.42			.38		.56
χ^2	13.68		14.25		2.22	
df	7		7		7	
p value	.057		.046		.947	
CFI	.98		1.0			
RMSEA	.086		0			
GFI	.97		.99			

[1] Pasteur results indicated a negative error variance for VISPERC.

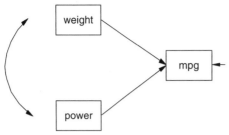

■ **Figure 8.3:** REGRESSION MODEL (AMOS)

the model fit and parameter estimates of each sample to determine whether they differ significantly. We therefore took two random samples without replacement from the *cars.sav* data.

The descriptive data for the two samples are in Table 8.4. A visual inspection shows that the correlations, means, and standard deviations look similar.

Table 8.4: Sample Descriptive Statistics

Sample 1 (N = 206)			
Correlation Matrix	mpg	weight	power
	1.0		
	−.82	1.0	
	−.778	.865	1.0
Means	23.94	2921.67	104.23
Standard Deviations	8.140	835.421	41.129
Sample 2 (N = 188)			
Correlation Matrix	mpg	weight	power
	1.0		
	−.823	1.0	
	−.760	.855	1.0
Means	23.59	2952.02	102.72
Standard Deviations	7.395	805.372	36.234

The global model fit, as well as the results for each sample from the analysis, are reported in Table 8.5 for the regression equation: mpg = weight + power. The multiple regression results for the complete data ($N = 394$) indicated what we expected for results in terms of R^2 values and regression coefficients. We can visually compare our two individual sample parameter estimates. The results appear to be very similar. Structural equation modeling software, however, provides the

Table 8.5: Multiple Sample Regression Model

Total Sample ($N = 394$)

Unstandardized Coefficients			Standardized Coefficients	t	p	R^2
	B	Std Error	B			.675
Constant	44.777	.825		54.307	.0001	
Vehicle Weight	−.005	.001	−.551	−9.818	.0001	
Horsepower	−.061	.011	−.299	−5.335	.0001	

Sample 1 ($N = 206$)

Unstandardized Coefficients			Standardized Coefficients	t	p	R^2
	B	Std Error	B			.692
Constant	46.214	1.193		38.723	.0001	
Vehicle Weight	−.006	.001	−.585	−7.550	.0001	
Horsepower	−.054	.015	−.272	−3.509	.0001	

Sample 2 ($N = 188$)

Unstandardized Coefficients			Standardized Coefficients	t	p	R^2
	B	Std Error	B			.689
Constant	45.412	1.166		38.957	.0001	
Vehicle Weight	−.006	.001	−.642	−8.114	.0001	
Horsepower	−.043	.016	−.212	−2.675	.0001	

capability of testing whether our results (parameter estimates) are statistically different.

SEM software provides the ability to compare both samples rather than having to run separate multiple regression programs on each sample and hand calculate a t test or z test for differences in the regression coefficients. The sample comparison yielded a non-significant chi-square ($\chi^2 = 2.01$, $df = 3$, $p = .57$), which indicates that the two samples do not have statistically different parameter estimates in the regression model. So, it seems reasonable to have a total sample regression coefficient of −.551 for *vehicle weight*, compared to the individual sample regression coefficients of −.585 and −.642, respectively. Looking at the regression coefficient for *horsepower* predicting *mpg,* we find individual regression coefficients of −.272 and −.212, respectively. So, it seems reasonable to have a common regression coefficient of −.299 for horsepower. Also, notice the R^2 values for each sample. We find that for each individual sample, the R^2 values were .692 and .689, respectively. So, once again, the total sample R^2 value of .675 is reasonable. Our interpretation would suggest that two-thirds of the miles per gallon variation for a car can be explained by a vehicle's weight and horsepower. The negative regression coefficients are expected because as weight and horsepower increase, miles per gallon should decrease.

SUMMARY

The SEM multiple group approach is useful for testing whether group differences exist for a theoretically specified model. The group comparisons can be made for regression, path, factor, or structural models. When the multiple group method is used in factor analysis, the purpose might be to establish measurement invariance, that is, the groups have the same construct or latent variable meaning. However, a researcher can hypothesize that group differences exist in factor models, as well as the other model types if the interest is in hypothesizing group differences. Multiple sample testing uses the same SEM approach, but now the focus is on whether the samples of data yield similar or different parameter estimates, or overall model fit, when comparing multiple regression, path models, confirmatory factor models, or structural equation models.

This chapter presented only one example for each of the applications because a more in-depth coverage is beyond the scope of this book. However, your SEM software program should provide other examples. An Internet search will also yield articles and software examples. We have placed the data and programs used in this chapter on the book website.

EXERCISES

1. Multiple Group Model

Create an SEM program that produces output to determine if path coefficients are statistically significantly different. You will need the separate data set information provided below to perform this task. Also provide a path diagram with interpretation of results.

The path model tests whether job satisfaction (satis) is indicated by boss attitude (boss) and the number of hours worked (hrs). The boss attitude (boss) is in turn indicated by the employee satisfaction (satis). The boss attitude (boss) is also indicated by the type of work performed (type), level of assistance provided (assist), and evaluation of the work (eval). There would be two model statements or equations specified as follows:

```
satis = boss hrs
boss = type assist eval satis
```

Note: Because a reciprocal relation exists between boss and satis, the errors would need to be correlated to obtain the correct path coefficients.

The data set information to be used to test hypotheses of equal or unequal parameter estimates in a path model between Germany and the United States are listed below.

Germany

```
Observed Variables satis boss hrs type assist eval
Sample Size 400
Means 1.12 2.42 10.34 4.00 54.13 12.65
Standard Deviation 1.25 2.50 3.94 2.91 9.32 2.01
Correlation Matrix
1.00
.55 1.00
.49 .42 1.00
.10 .35 .08 1.00
.04 .46 .18 .14 1.00
.01 .43 .05 .19 .17 1.00
```

United States

```
Observed Variables satis boss hrs type assist eval
Sample Size = 400
Means: 1.10 2.44 8.65 5.00 61.91 12.59
Standard Deviations: 1.16 2.49 4.04 4.41 4.32 1.97
Correlation Matrix
1.00
.69 1.00
.48 .35 1.00
.02 .24 .11 1.00
.11 .19 .16 .31 1.00
.10 .28 .13 .26 .18 1.00
```

2. Multiple Sample Model

Nursing programs are interested in knowing if their outcomes are similar from one semester to the next. Two semesters of data were obtained on how student effort (effort) and learning environment (learn) predicted clinical competence (comp) in nursing. The regression model is: comp = effort + learn.

Chapter 9

SECOND-ORDER CFA MODELS

CHAPTER CONCEPTS

Second-order factor model
Model specification
Model identification
Model estimation
Model testing
Model interpretation

In this chapter we expand our understanding of measurement models. Measurement models which define the latent variables used in structural models will require most of your time, knowledge, and skill. Model specification is the first important step in establishing a measurement model. If the indicator variables for the latent variables are not properly selected and specified based on prior theory and research, then the measurement model will not adequately reflect the variable relations in the covariance matrix or yield a valid latent variable construct.

When the measurement model is not specified correctly, which often happens when too many parameter estimates are indicated for the number of variables, model identification fails. Model identification requires that degrees of freedom be equal to or greater than 1 ($df \geq 1$). SEM software programs show the degrees of freedom, and generally report when it is zero (just-identified model) or negative (under-identified model).

Model estimation is also problematic in that continuous variables, ordinal variables, or a mixture of both are used in the measurement model. When this occurs a different sample matrix (asymptotic covariance matrix) and estimation method (diagonally weighted least squares) are usually required to achieve robust estimates of parameter estimates. LISREL9, using raw data, will automatically perform robust estimation of standard errors and chi-square goodness-of-fit measures under non-normality, and routinely provide the asymptotic matrix and robust chi-square statistic. In the presence of missing data, the FIML (full information maximum likelihood) estimation method will be used. All of this works for both continuous and ordinal variables. Mplus also provides correct matrices, automated start values, and different estimation methods for categorical (ordinal) and continuous variables.

Model testing involves reporting the chi-square, degrees of freedom, and p value. The chi-square statistic is sensitive to sample size, with large sample sizes over-reporting significance and small sample sizes under-reporting significance. Consequently, other fit indices (although not statistically based) are used to substantiate model fit, model comparison, or model parsimony. These other fit indices generally use the chi-square value, degrees of freedom, and sample size in their calculations. We have also suggested that parameter estimates be examined for statistical significance. It is possible to have a global chi-square value that is non-significant, thus indicating a good data to model fit, yet one of the parameter estimates may not be statistically significant. Remember that the chi-square statistic is a test of the overall model fit.

Please be aware that our discussion will only scratch the surface of the many exciting new developments in structural equation modeling related to measurement models. Some of these new applications have been included in chapters of books (Marcoulides & Schumacker, 1996, 2001; Schumacker & Marcoulides, 1998) and journal articles. In addition, the newest versions of SEM software, for example, LISREL9, have included these capabilities with software examples and further explanations. Please check your SEM software for any new features. Our intention is to provide a basic understanding of these topics to further your interest in the structural equation modeling approach.

MODEL SPECIFICATION

A second-order factor model is specified where the three first-order factors explain a higher-order factor structure. Theory plays an important role in justifying a higher-order factor. *Visual*, *verbal*, and *speed* are three psychological factors that most likely indicate a second-order factor, namely *Ability*. An ability second-order factor model is therefore hypothesized and diagrammed in Figure 9.1 using data from Jöreskog and Sörbom (1993).

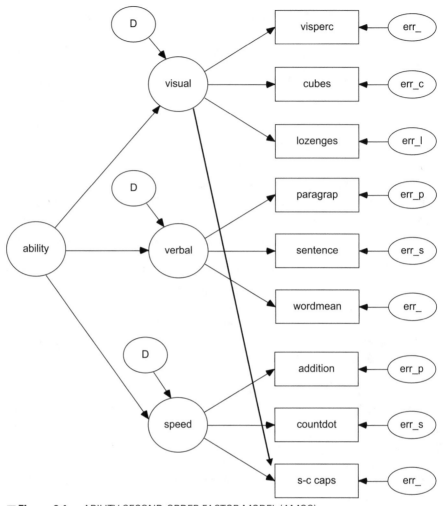

■ **Figure 9.1:**   ABILITY SECOND-ORDER FACTOR MODEL (AMOS)

MODEL IDENTIFICATION

Recall from Chapter 7 that the number of distinct values in the S matrix (sample covariance matrix) is $p(p + 1)/2$. For the 9 indicator variables in the second-order factor model: $p(p + 1)/2 = 9(10)/2 = 45$ distinct values (variances and covariances). The number of free parameters (saturated model – all paths specified) is $p(p + 3)/2$. For the saturated model: $p(p + 3)/2 = 9(12)/2 = 54$ total free parameters in the model that could be estimated. Consequently, if we estimated all parameters in the model, we would have a negative degrees of freedom ($df = 45 - 54 = -9$), and the model would not be identified. The number of parameters to be estimated in the model must be less than 45 to have a positive degrees of freedom ($df \geq 1$).

The second-order factor model has degrees of freedom = 23. The parameters to be estimated in the model are 10 factor loadings, 3 second-order factor loadings, and 9 error variances (22 parameters). The degrees of freedom for the model is computed as $45 - 22 = 23$. The model is over-identified, that is, fewer parameter estimates than distinct values in the matrix, which is what we desire in SEM analyses. A just-identified model would have the number of parameters estimated equal to the number of distinct values, thus $df = 45 - 45 = 0$. An under-identified model would have more parameters estimated than distinct values, thus a negative degrees of freedom, $df = 45 - 54 = -9$.

MODEL ESTIMATION

The sample data are nine psychological variables that identified three common factors (*Visual*, *Verbal*, and *Speed*). The second-order factor model hypothesized that these three common factors indicate a higher-order second factor, *Ability*. We will be using a correlation matrix, rather than a variance–covariance matrix, so the variables have mean = 0 and standard deviation = 1. Essentially, this places the variables on the same scale of measurement, but we run the risk of the standard errors being inflated. If the standard errors are inflated, then a parameter estimate may not be statistically significant or biased (recall, t or z = parameter estimate divided by standard error). The maximum likelihood estimation method was used (default value) with the assumption of multivariate normality.

MODEL TESTING

The second-order factor model is now ready to be tested for model fit. The second-order factor model is based on an example in LISREL9, *SPLEX* folder (EX5A.spl). The second-order factor model includes the *Ability* latent variable and sets the variance of this higher-order second factor to 1.0. We have changed this model to include "*S-C CAPS*" loading on two latent variables, *Visual* and *Speed* (Figure 9.1). The *selected* model fit indices indicated that the hypothesized second-order factor model has an acceptable fit ($\chi^2 = 28.744$, $p = .189$, $df = 23$; RMSEA = .04; GFI = .958).

MODEL INTERPRETATION

The measurement equations with unstandardized factor loadings and standard errors indicated statistically significant parameter estimates. There are eight measurement equations for indicators of a single factor, and one equation with an indicator ("SC-CAPS") for the two latent variables. The computer program output is shown as follows:

```
Measurement Equations (unstandardized)

    VIS PERC = 0.708*Visual, Errorvar.= 0.498 , R² = 0.502
    Standerr              (0.0899)
    Z-values                5.546
    P-values                0.000


    CUBES = 0.483*Visual, Errorvar.= 0.766, R² = 0.234
    Standerr (0.102)      (0.100)
    Z-values  4.726         7.644
    P-values  0.000         0.000


    LOZENGES = 0.650*Visual, Errorvar.= 0.578 , R² = 0.422
    Standerr(0.110)        (0.0908)
    Z-values 5.905          6.361
    P-values 0.000          0.000


    PAR COMP = 0.868*Verbal, Errorvar.= 0.247 , R² = 0.753
    Standerr              (0.0511)
    Z-values                4.825
    P-values                0.000


    SEN COMP = 0.830*Verbal, Errorvar.= 0.311 , R² = 0.689
    Standerr (0.0723)     (0.0535)
    Z-values 11.481         5.820
    P-values  0.000         0.000


    WORDMEAN = 0.825*Verbal, Errorvar.= 0.319 , R² = 0.681
    Standerr (0.0723)     (0.0539)
    Z-values 11.412         5.929
    P-values  0.000         0.000


    ADDITION = 0.681*Speed, Errorvar.= 0.536 , R² = 0.464
    Standerr              (0.0928)
    Z-values                5.778
    P-values                0.000


    COUNTDOT = 0.859*Speed, Errorvar.= 0.262, R² = 0.738
    Standerr (0.147)      (0.113)
    Z-values  5.848         2.320
    P-values  0.000         0.020
```

Table 9.1: Second-order Factor Model (Standardized Solution)

Indicator Variables	Visual	Verbal	Speed
VIS PERC	.708		
CUBES	.483		
LOZENGES	.650		
PAR COMP		.868	
SEN COMP		.830	
WORDMEAN		.825	
ADDITION			.681
COUNTDOT			.859
S-C CAPS	.457		.419

```
S-C CAPS = 0.457*Visual + 0.419*Speed, Errorvar.= 0.467 , R² = 0.533
Standerr  (0.101)      (0.0931)              (0.0724)
Z-values  4.519        4.499                 6.446
P-values  0.000        0.000                 0.000
```

The measurement model results can also be indicated as standardized output in SEM programs. This permits a comparison of the factor loadings. Table 9.1 displays the standardized factor loadings for the three latent variables. The factor loadings can be considered *validity coefficients*, that is, they measure how much the indicator variable defines the latent variable. Ideally, the factor loadings should be .60 (60%) or higher, so the CUBES factor loading is lower than desired. The "SC-CAPS" indicator variable shares two latent variables, so we would combine these for our communality estimate, thus these values being lower than .60 is not necessarily undesirable. The factor loadings of the three indicator variables (PAR COMP; SEN COMP; WORDMEAN) for the *Verbal* latent variable are desirable. When we average the standardized factor loadings, it indicates how much factor variance is explained. For example: (.868 + .830 + .825)/3 = .841, so 84% of the *Verbal* factor variance is explained (16% is unexplained).

The structural equations in the output indicate the strength of relationship between the first-order factors and the second-order factor, *Ability*. *Visual* (.987) is indicated as a stronger measure of *Ability*, followed by *Verbal* (.565) and *Speed* (.395); with all three being statistically significant ($z > 1.96$). Therefore, *Ability* is predominantly a function of visual perception of geometric configurations with complementary verbal skills and speed in completing numerical tasks. The computer output is shown as follows:

Structural Equations

```
Visual = 0.987*Ability, Errorvar.= 0.0257, R² = 0.974
Standerr   (0.228)              (0.400)
Z-values   4.324                0.0643
P-values   0.000                0.949


  Verbal = 0.565*Ability, Errorvar.= 0.681, R² = 0.319
Standerr   (0.140)              (0.170)
Z-values   4.029                4.010
P-values   0.000                0.000


  Speed = 0.395*Ability, Errorvar.= 0.844, R² = 0.156
Standerr   (0.131)              (0.226)
Z-values   3.010                3.729
P-values   0.003                0.000
```

SUMMARY

The second-order factor model permits other factors to represent a higher-order construct. We first examined the indicators for the first-order factors, which had statistically significant factor loadings. We then examined the second-order *Ability* factor, which was defined by *Visual*, *Verbal*, and *Speed* first-order factors. We examined the factor loadings for these three first-order factors, and found them to be statistically significant indicators of *Ability*. The second-order factor model had a good data to model fit.

We hope that our discussion of this SEM application has provided you with a basic overview and introduction to second-order factor methods. We encourage you to read the references provided at the end of the chapter and run the exercise provided in the chapter. We further hope that the basic introduction in this chapter will permit you to read the research literature and better understand second-order factor models, which should support various theoretical perspectives. Attempting a few basic models will help you better understand the approach; afterwards, you may wish to conduct this SEM application in your own research.

EXERCISE

The psychological research literature tends to suggest that drug use and depression are leading indicators of suicide among teenagers. (*Note:* Set variance of Suicide = 1 for model identification purposes.) Given the following data set information, create and run an SEM program to conduct a second-order factor analysis. The basic program setup is shown as follows without the model equations:

```
Suicide second order factor analysis
Observed Variables: drug1 drug2 drug3 drug4 depress1 depress2
depress3 depress4
Sample Size 200
Correlation Matrix
 1.000
 0.628 1.000
 0.623 0.646 1.000
 0.542 0.656 0.626 1.000
 0.496 0.557 0.579 0.640 1.000
 0.374 0.392 0.425 0.451 0.590 1.000
 0.406 0.439 0.446 0.444 0.668  .488 1.000
 0.489 0.510 0.522 0.467 0.643  .591  .612 1.000
Means  1.879 1.696 1.797 2.198 2.043 1.029 1.947  2.024
Standard Deviations1.379 1.314 1.288 1.388 1.405  1.269 1.435
1.423
Latent Variables: drugs depress Suicide
```

The suicide second-order factor model is displayed in Figure 9.2. This provides the basis for writing the model equations in the second-order factor analysis program.

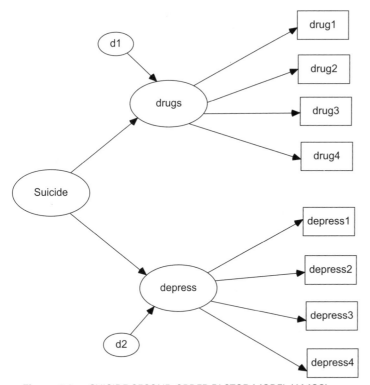

Figure 9.2:   SUICIDE SECOND-ORDER FACTOR MODEL (AMOS)

SUGGESTED READINGS

Chan, D. W. (2006, Fall). Perceived multiple intelligences among male and female Chinese gifted students in Hong Kong: The structure of the student multiple intelligences profile. *The Gifted Child Quarterly*, 50(4), 325–338.

Cheung, D. (2000). Evidence of a single second-order factor in student ratings of teaching effectiveness. *Structural Equation Modeling: A Multidisciplinary Journal*, 7, 442–460.

Marsh, H. W., & Hocevar, D. (1985). Application of confirmatory factor analysis to the study of self-concept: First- and higher order factor models and their invariance across groups. *Psychological Bulletin*, 97(3), 562–582.

Rand, D., Conger, R. D., Patterson, G. R., & Ge, X. (1995). It takes two to replicate: A mediational model for the impact of parents' stress on adolescent adjustment. *Child Development*, 66(1), 80–97.

REFERENCES

Jöreskog, K. G., & Sörbom, D. (1993). *LISREL 8: Structural equation modeling with the SIMPLIS command language*. Chicago: Scientific Software International.

Marcoulides, G., & Schumacker, R. E. (Eds.). (1996). *Advanced structural equation modeling: Issues and techniques*. Mahwah, NJ: Lawrence Erlbaum Associates.

Marcoulides, G., & Schumacker, R. E. (Eds.). (2001). *New developments and techniques in structural equation modeling*. Mahwah, NJ: Lawrence Erlbaum Associates.

Schumacker, R. E., & Marcoulides, G. A. (1998). *Interaction and nonlinear effects in structural equation modeling*. Mahwah, NJ: Lawrence Erlbaum.

## Chapter 10

DYNAMIC FACTOR MODEL

CHAPTER CONCEPTS

Dynamic factor model
Model specification
Model identification
Model estimation
Model testing
Model interpretation

A class of SEM applications that involve stationary and non-stationary latent variables across time with lagged (correlated) measurement error has been called dynamic factor analysis (Hershberger, Molenaar, & Corneal, 1996). A characteristic of the SEM dynamic factor model is that the same measurement instruments are administered to the same subject on two or more occasions. The purpose of the analysis is to assess change in the latent variable between the ordered occasions due to some event or treatment. When the same measurement instruments are used over two or more occasions, there is a tendency for the measurement errors to correlate (autocorrelation). For example, a specific sequence of correlated error, where error at Time 1 correlates with error at Time 2, and error at Time 2 correlates with error at Time 3, is called an ARIMA model in econometrics (Wheaton, Muthén, Alwin, & Summers, 1977).

MODEL SPECIFICATION

Educational research has indicated that anxiety increases the level of student achievement and performance. Psychological research in contrast indicates that anxiety has a negative effect upon individuals, and thus should interfere with or have a decreasing impact on the level of achievement and performance. Is it possible that both academic areas of research are correct?

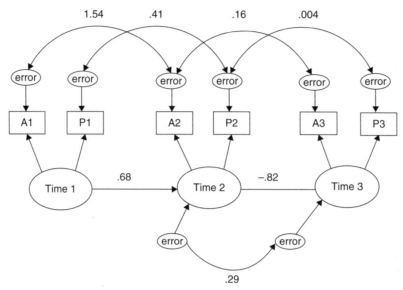

■ **Figure 10.2:** DYNAMIC FACTOR MODEL (STANDARDIZED SOLUTION)

SUMMARY

The dynamic factor model extends the repeated measures design to include latent variables. When using latent variables, more than one observed indicator variable can be used. The dynamic factor model presented in this chapter included two indicator variables. The two observed variables were measured at each of three time points by the same subjects. In addition, the time intervals were the same. It is possible to have time-varying measurements such that they are not at the same interval of measurement. You will find many more examples of dynamic factor modeling in the macro-economic and econometric areas. In education and psychology the application has centered more on the latent growth model, which we cover in Chapter 14.

EXERCISE

A sports physician was interested in studying heart rate and muscle fatigue of female soccer players. She collected data after three soccer games over a three-week period. A dynamic factor model was used to determine if heart rate and muscle fatigue were stable across time for the 150 female soccer players.

Create an SEM program to analyze and interpret the dynamic factor model. Include a diagram of the dynamic factor model. The data set information including observed variables, covariance matrix, sample size, and latent variables are listed below without the model equations.

```
Observed Variables: HR1 MF1 HR2 MF2 HR3 MF3
Covariance Matrix
  10.75
   7.00    9.34
   7.00    5.00   11.50
   5.03    5.00    7.49    9.96
   3.89    4.00    3.84    3.65    9.51
   2.90    2.00    2.15    2.88    3.55    5.50
Sample Size: 150
Latent Variables: Time1 Time2 Time3
```

SUGGESTED READINGS

Chow, S. M., Nesselrade, J. R., Shifren, K., & McArdle, J. J. (2004). Dynamic structure of emotions among individuals with Parkinson's disease. *Structural Equation Modeling, 11*(4), 560–582.

Kroonenberg, P. M., van Dam, M., van Uzendoorn, M. H., & Mooijaart, A. (1997, May). Dynamics of behaviour in the strange situation: A structural equation approach. *British Journal of Psychology, 88,* 311–332.

Zuur, A. F., Fryer, R. J., Jolliffe, I. T., Dekker, R., & Beukema, J. J. (2003). Estimating common trends in multivariate time series using dynamic factor analysis. *Environmetrics, 14*(7), 665–685.

REFERENCES

Hershberger, S. L., Molenaar, P. C. M., & Corneal, S. E. (1996). A hierarchy of univariate and multivariate structural times series models. In Marcoulides, G., & Schumacker, R. E. (Eds.). *Advanced structural equation modeling: Issues and techniques* (pp. 159–194). Mahwah, NJ: Lawrence Erlbaum Associates.

Wheaton, B., Muthén, B., Alwin, D., & Summers, G. (1977). Assessing reliability and stability in panel models. In D. R. Heise (Ed.), *Sociological Methodology* (pp. 84–136). San Francisco: Jossey-Bass.

Chapter 11

MULTIPLE-INDICATOR MULTIPLE-CAUSE (MIMIC) MODEL

CHAPTER CONCEPTS

MIMIC model
Model specification
Model identification
Model estimation
Model testing
Model modification
Model interpretation

MODEL SPECIFICATION

The term MIMIC refers to multiple indicators and multiple causes, and defines a particular type of SEM model. The MIMIC model involves using latent variables that are predicted by observed variables. We discuss an example by Jöreskog and Sörbom (1996a; 1996b, example 5.4, pp. 185–187). The MIMIC model shows a latent variable, social participation (*social*) that is defined by church attendance (*church*), memberships (*member*), and friends (*friends*). The social participation latent variable is predicted by the observed variables, *income*, *occupation*, and *education*. The MIMIC model for the social participation latent variable is diagrammed in Figure 11.1.

The MIMIC model indicates a latent variable, *social*, which has arrows pointing out to the three observed indicator variables (church, member, friends) with separate measurement error terms for each. This is the measurement part of the MIMIC model that defines the latent variable. In the MIMIC model, the latent variable, *social*, also has arrows pointing toward it from the three observed

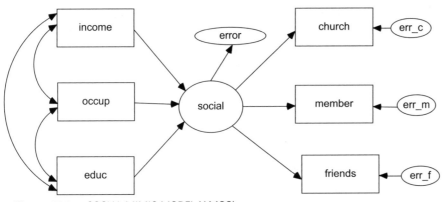

■ **Figure 11.1:**   SOCIAL MIMIC MODEL (AMOS)

predictor variables, which have implied correlations among them (curved arrows). This is the structural part of the MIMIC model that uses observed variables to predict a latent variable. The MIMIC model diagram also shows the prediction error for the latent variable, *social*. The important thing to remember in a MIMIC model is the direction of the arrows.

MODEL IDENTIFICATION

Model identification pertains to whether the estimates in the MIMIC model can be calculated, which is quickly gauged by the degrees of freedom. Do you recall how the degrees of freedom are determined? There are a total of 15 free parameters to be estimated in the MIMIC model. The number of distinct values in the variance–covariance matrix, S, based on 6 observed variables is: $p(p + 1)/2 = 6(6 + 1)/2 = 21$. The degrees of freedom are computed by subtracting the number of free parameters from the number of distinct parameters in the matrix S, which is $21 - 15 = 6$. The model is considered identified given the positive degrees of freedom.

MODEL ESTIMATION

The MIMIC model diagram provides the basis for specifying the SEM program, and determining whether the model is identified (positive degrees of freedom). The SEM program will need to include the observed variables, sample size, correlation matrix (standardized variables), or covariance matrix, and the equations that reflect the MIMIC model to permit estimation of parameters in the model.

Model estimation involves selecting the estimation method for the parameters in the MIMIC model. The SEM program default is usually maximum likelihood estimation, which may or may not be your choice. The SEM MIMIC model goodness-of-fit criteria follow from the chi-square computed to determine whether a reasonably good fit of the data to the MIMIC model exists. We are only reporting the chi-square, RMSEA, and GFI values for determining model fit.

MODEL TESTING

The $\chi^2 = 12.04$, $df = 6$, and $p = .061$, which suggests a reasonably good fit of the data to the MIMIC model. The goodness-of-fit index (GFI) suggested that 99% of the variance–covariance in matrix S is reproduced by the MIMIC model. The standardized solution indicated factor loadings of .47 (church), .74 (member), and .40 (friends). However, the z value in the computer output dropped *church* as an important indicator variable in defining the latent variable, *social*. The observed variables, *member* ($z = 6.718$) and *friends* ($z = 6.035$), were therefore selected to define the latent variable *social*. The measurement equations from the computer output are listed below.

Measurement Equations

```
church = 0.466*social, Errorvar.= 0.783, R² = 0.217
Standerr                (0.0575)
Z-values                13.621
P-values                 0.000

 member = 0.735*social, Errorvar.= 0.459, R² = 0.541
Standerr (0.109)               (0.0753)
Z-values  6.718                 6.102
P-values  0.000                 0.000

 friends = 0.402*social, Errorvar.= 0.839, R² = 0.161
Standerr (0.0665)              (0.0577)
Z-values  6.035                14.526
P-values  0.000                 0.000
```

Note: Because a matrix was used rather than raw data, standard error and z value are not output for the reference indicator variable, *church*. The *HELP* menu offers this explanation: LISREL for Windows uses a reference indicator (indicator with a unit factor loading) to set the scale of each of the endogenous latent variables of the model. If you do not specify reference indicators for the endogenous latent variables of your model, it will select a reference indicator for each endogenous latent variable of your model. Although it scales the factor loadings to obtain the appropriate estimates for the factor loadings of

the reference indicators, it does not use the delta method to compute the corresponding standard error estimates.

The observed independent variables (income, occup, and educ) in the MIMIC model were correlated amongst themselves as identified in the correlation matrix of the SEM program output:

```
income occup educ
 1.000
 .304 1.000
 .305 .344 1.000
```

The structural equation indicated that the latent variable *social* had 26% of its variance predicted ($R^2 = .26$), with 74% unexplained error variance due to random or systematic error, and variables not in the MIMIC model. The z values for the structural equation coefficients indicated that *occup* (occupation) didn't statistically significantly predict *social* (z = parameter estimate divided by standard error = .097/.056 = 1.73 is less than z = 1.96 at the .05 level of significance, two-tailed test, whereas *income* (z = 3.82) and *educ* (z = 4.93) were statistically significant at the .05 level of significance). The structural equation with coefficients, standard errors in parentheses, and associated z values are listed below.

Structural Equation

```
social = 0.232*income + 0.0973*occup + 0.334*educ, Errorvar.= 0.742, R² = 0.258
Standerr  (0.0608)      (0.0563)       (0.0675)       (0.170)
Z-values   3.821         1.728          4.938          4.353
P-values   0.000         0.084          0.000          0.000
```

MODEL MODIFICATION

The original MIMIC model was modified by dropping *church* as an indicator variable, and *occup* as a predictor variable. The MIMIC model diagram with these modifications appears in Figure 11.2. The model modification fit criteria are more acceptable, indicating an almost perfect fit of the data to the MIMIC model, because the Minimum Fit Function χ^2 value was close to zero.

The modified social MIMIC model indicated a good data to model fit ($\chi^2 = .19$, $df = 1$, and $p = .66$). This modified model reflects dropping both a non-significant factor loading in the measurement model (*church*), and a non-significant

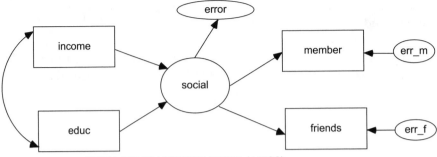

■ **Figure 11.2:**   SOCIAL MIMIC MODIFIED MODEL (AMOS)

predictor variable (*occup*) in the structural model. The computer output is now shown as:

```
Measurement Equations

   member = 0.630*social, Errorvar.= 0.603, R² = 0.397
Standerr                  (0.0822)
Z-values                  7.332
P-values                  0.000

   friends = 0.420*social, Errorvar.= 0.823, R² = 0.177
Standerr (0.0757)               (0.0602)
Z-values  5.553                 13.677
P-values  0.000                 0.000
```

Note: Because a matrix was used rather than raw data, standard errors are not output for one of the reference indicator variables, *member* = 0.63 (*social*). The *HELP* menu offers further explanation as noted above.

The structural equation now indicated two statistically significant predictor variables with R^2 = .36. This also implies that 64% of the latent variable variance is left unexplained, mostly due to random or systematic error or other variables not included in the MIMIC model. Ideally we desire a higher R^2 value to indicate more explanation of the variability in the latent variable, *social*. The computer output is shown as:

```
Structural Equations

   social = 0.313*income + 0.424*educ, Errorvar.= 0.641, R² = 0.359
Standerr (0.0625)    (0.0638)          (0.189)
Z-values  5.014       6.653             3.393
P-values  0.000       0.000             0.001
```

SUMMARY

MIMIC models permit the specification of one or more latent variables in a measurement model with one or more observed variables as predictors of the latent variables in a structural model. This type of SEM model demonstrates how observed variables can be incorporated into a theoretical model and tested. In some cases, the researcher first establishes a good fit of the data to the measurement model, so a latent variable is well defined. Afterwards, it is straightforward to include observed predictor variables of the latent variable(s). We followed the five basic steps in the SEM: model specification, model identification, model estimation, model testing, and model modification to obtain our best data to model fit. We interpreted both the model fit (chi-square test), as well as the statistical significance of the parameter estimates (z values).

It is important in MIMIC models to understand the direction of the paths in the theoretical model. This corresponds to how the equations are written in an SEM program. For illustration, we provide the following explanation. The measurement model statement has the indicator variables on the left-hand side of the equation (member, friends) when defining the latent variable *social*. The structural model has the observed predictor variables on the right-hand side of the equation (income, educ), which predict the dependent latent variable, *social*. The structural model statement is similar to how we write our prediction equations. If not done correctly, the path diagram should reveal the error in the drawing of the model.

EXERCISE

Create and run an SEM program given the MIMIC model in Figure 11.3. The MIMIC model includes the latent variable *job satisfaction* (satisfac), which is defined by two observed variables: *peer rating* and *self rating*. A person's income level (income), which shift they worked (shift), and age (age) are observed predictor variables of *job satisfaction*.

Interpret the results including any model modification, significance of coefficients, and R^2 value. The data set information is:

```
Observed Variables peer self income shift age
Sample Size 530
Correlation Matrix
1.00
 .42 1.00
 .24 .35 1.00
 .13 .37 .25 1.00
 .33 .51 .66 .20 1.00
```

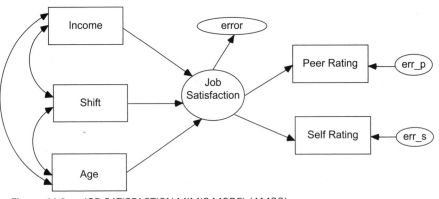

▪ **Figure 11.3:**   JOB SATISFACTION MIMIC MODEL (AMOS)

SUGGESTED READINGS

Anderson, K. G., Smith, G. T., & McCarthy, D. M. (2005). Elementary school drinking: The role of temperament and learning. *Psychology of Addictive Behaviors*, 19(1), 21–27.

Sanchez-Perez, M., & Iniesta-Bonillo, M. A. (2004, Winter). Consumers felt commitment towards retailers: Index development and validation. *Journal of Business and Psychology*, 19(2), 141–159.

Shenzad, S. (2006). The determinants of child health in Pakistan: An economic analysis. *Social Indicators Research*, 78, 531–556.

REFERENCES

Jöreskog, K., & Sörbom, D. (1996a). *LISREL 8: User's reference guide*. Chicago, IL: Scientific Software International.

Jöreskog, K., & Sörbom, D. (1996b). *LISREL 8: Structural equation modeling with the SIMPLIS command language*. Chicago, IL: Scientific Software International.

## Chapter 12

MIXED VARIABLE AND MIXTURE MODELS

CHAPTER CONCEPTS

Mixed variable model–different variable types
Model specification
Model identification
Model estimation
Model testing
Model modification
Robust chi-square statistic
Mixture model–latent classes

MIXED VARIABLE MODEL—DIFFERENT VARIABLE TYPES

SEM mixture models include both categorical and continuous observed variables. SEM was originally created using continuous variables in a sample variance–covariance matrix (Pearson correlation matrix with means and standard deviations); however, today SEM models with nominal, ordinal, interval, and ratio-level observed variables can be used in SEM. For example, Mplus and LISREL provide numerous examples using a set of mixed variable types where parameter estimates are computed with appropriate estimation methods.

The use of both categorical and continuous variables, however, requires using another type of matrix than the Pearson correlation matrix and associated variance–covariance matrix in SEM programs. In the LISREL9 software program (Jöreskog, Sörbom, du Toit, & du Toit, 2001), raw data input will produce the type of matrix needed for the different variable types in the model. SEM software programs also provide ways to define variables, for example, in PRELIS, they are defined by the **CO** command (by default the variable must have a minimum of 15 categories), the **OR** command for ordinal variables, or the **CL** command for class or group variables. Mplus uses the commands *Categorical Are* for binary and ordinal variables, *Nominal Are* for true dichotomy variables, and *Grouping Is* for group membership variable.

SEM software programs generally output normal theory variance–covariance matrices (correlation between continuous variables), polychoric matrices (correlation between ordered categorical variables), polyserial matrices (correlation between continuous and ordinal variables), asymptotic variance–covariance matrices (continuous and/or ordinal variables with non-normality), and augmented moment matrices (matrices with variable means). Consequently, in SEM, we must pay attention to the type of observed variables and the associated type of matrix that needs to be used given the theoretical model. The model-fit statistic in these types of models becomes a robust model-fit measure where the asymptotic covariance matrix with maximum likelihood estimation is required to obtain the Satorra–Bentler robust χ^2 statistic.

MODEL SPECIFICATION

The mixed variable model example uses observed variables from an SPSS data set, *bankloan.sav*. This is a hypothetical data set that concerns a bank's efforts to reduce the rate of loan defaults. The file contains financial and demographic information on 850 past and prospective customers. The data set is located in the IBM SPSS *Samples* folder, and available on the book website.

A theoretical model was hypothesized that financial *Ability* was a predictor of *Debt*. The observed variables age, level of education, years with current employer, years at current address, and household income in thousands were used as indicators of the latent independent variable, *Ability*. The observed variables credit card debt in thousands and other debt in thousands were used as indicators of the latent dependent variable, *Debt*.

You will need to import the SPSS data file (*bankloan.sav*) into the SEM software and save it as the required file type (for example, in LISREL9 it would be *bankloan.lsf*; in Mplus it would be *bankloan.txt*). Once you have saved the file, delete the variables *DEBTINC*, *Default*, *preddef1*, *preddef2*, and *preddef3*, leaving seven variables for the theoretical mixed variable model. We decided that these five variables (*DEBTINC*, Default, preddef1, preddef2, and preddef3) were not good indicators in our theoretical model. The level of education (ED) variable was ordinal (OR), while all other variables were considered continuous (CO). The model also indicates that *Ability* predicts *Debt*, and is diagrammed in Figure 12.1.

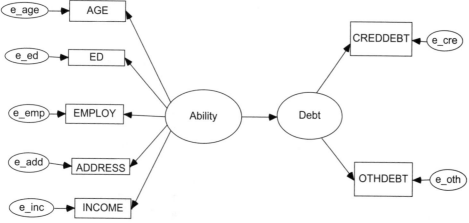

■**Figure 12.1:** ABILITY MIXED VARIABLE MODEL (AMOS)

MODEL IDENTIFICATION

The model has five indicators of the latent variable *Ability*. We desire the factor loadings (validity coefficients) to be statistically significant and sufficiently high in value to explain the factor variance in the latent variable. There are two indicator variables for the latent variable *Debt*. Once again we desire these factor loadings to be statistically significant. The independent latent variable (*Ability*) is predicting the dependent latent variable (*Debt*) in the structural model. Together, the 7 variables in the matrix equal 28 distinct values, $p(p + 1)/2 = 28$, or 7 variances and 21 covariance terms. The theoretical model indicates parameters estimates for 7 factor loadings, 7 error terms, and 1 structural coefficient, which is 15. The degrees of freedom is $28 - 15 = 13$. We therefore consider the model identified.

In addition to the degrees of freedom being positive, we also desire the sample variance–covariance matrix to not display any problems. For example, we desire a positive determinant of the matrix, which is indicated as the generalized variance. The determinant of a matrix is equal to the product of its eigenvalues, that is, $det(A) = \lambda_1\lambda_2\lambda_3....\lambda_n$. We desire positive eigenvalues because they indicate individual variable variance. The eigenvectors associated with each eigenvalue are used to solve the set of simultaneous equations, and thus used to compute the parameter estimates. LISREL9 now displays these values for the researcher. Other SEM programs display similar values, especially the determinant of the matrix and condition code. If you see error messages related to these values, stop! You have a problem with the sample variance–covariance matrix, probably related to multicollinearity or linear dependency amongst the observed variables. In our example, these values were positive, so along with the degrees of freedom, the model is identified. This means we can calculate parameter estimates, and with the standard errors, compute their statistical significance.

MODEL ESTIMATION

We need to enter the sample data and output a polyserial correlation matrix, *bankloan.mat,* using an SEM program. Some SEM programs simply require a command statement to indicate the different matrix when using raw data. The PRELIS program uses a DA command to specify the 7 input variables (NI = 7) with 850 observations (NO = 850); missing data are identified by a zero (MI = 0) and we treat missing data listwise (TR = LI). The SY command identifies the PRELIS system file (*bankloan.lis*) that contains the raw data. The OU command identifies the type of matrix to be computed—that is, polyserial matrix (MA = PM), and the name of the polyserial matrix (PM = *bankloan.mat*). PRELIS can output eight different types of matrices (Jöreskog & Sörbom, 1996, pp. 92–93), so it is important to use the correct type of matrix when conducting SEM analyses.

The following PRELIS program was saved as *bankloan.prl*, and run to create the required matrix for the SEM program (Jöreskog & Sörbom, 1996). The OU command line specifies the type of correlation matrix (MA = PM; polyserial matrix), where the PM command lists the name of the file.

```
Polyserial correlation matrix
DA NI = 7 NO=850 MI = 0 TR = LI
SY FI = bankloan.lsf
CO AGE
OR ED
CO EMPLOY
CO ADDRESS
CO INCOME
CO CREDDEBT
CO OTHDEBT
OU MA = PM PM = bankloan.mat
```

The polyserial correlation matrix, *bankloan.mat*, can now be used to obtain parameter estimates in the mixed variable model.

MODEL TESTING

The sample data for the model had two variables, *EMPLOY* and *ADDRESS*, with missing data, which left an effective sample size of $N = 723$. The resulting saved polyserial correlation matrix, *bankloan.mat*, was used in our model to test model fit and the statistical significance of the parameter estimates. The means and standard deviations were also input into the program. Notice that ED is coded 0 and 1, respectively, for mean and standard deviation. The SIMPLIS program (*bankloan.spl*) for the mixed variable model would be:

```
Mixture Model using Polyserial Correlation Matrix
Observed Variables: AGE ED EMPLOY ADDRESS INCOME CREDDEBT OTHDEBT
Sample Size 723
Correlation Matrix from file bankloan.mat
Means: 35.903 0.000 9.593 9.216 49.732 1.665 3.271
Standard Deviations: 7.766 1.000 6.588 6.729 40.243 2.227 3.541
Latent Variables Ability Debt
Relationships
AGE ED EMPLOY ADDRESS INCOME = Ability
CREDDEBT OTHDEBT = Debt
Debt = Ability
Number of Decimals = 3
Path Diagram
End of Problem
```

The SIMPLIS program contains command lines that are easily converted to the syntax of other SEM programs. The basic command lines for *Title*, *Observed Variables*, *Sample Size*, *Correlation Matrix*, *Means*, and *Standard Deviations* can be found in other SEM programs. The model statements using *Latent Variables* and *Relationships* define the measurement model and structural model in Figure 12.1. The measurement models for the two latent variables are specified in the command lines:

```
AGE ED EMPLOY ADDRESS INCOME = Ability
CREDDEBT OTHDEBT = Debt
```

The observed variables indicate the two latent variables (Ability; Debt), which are named on the *Latent Variables* command line. The structural model is given in the command line:

```
Debt = Ability
```

This indicates that *Debt* is predicted by *Ability*. The measurement models create the latent variables, which are then used in the structural equation. Thus, *Ability* latent scores (independent latent variable) are predicting *Debt* latent scores (dependent latent variable). SEM programs today also output a diagram of the model (except for AMOS, which inputs the diagram, and then writes the model statements).

The maximum likelihood chi-square model fit results were **not** adequate ($\chi^2 = 436.32$, $df = 13$, $p > 0.00001$, RMSEA $= 0.212$), so we don't bother with interpreting the parameter estimates for statistical significance. The computer output also indicated a condition code with an error message for severe multicollinearity, so one or more variables may be redundant or linearly related to another. We should therefore examine the observed variable correlations. The polyserial correlations in the file *bankload.mat* do not display the high intercorrelation associated with multicollinearity (Table 12.1). The issue most likely is linear dependency amongst the observed variables, hence the error message. We will proceed with modifying the model, which can be done by examining the modification indices to determine any substantive model modifications.

Table 12.1: Polyserial Correlation Matrix *(bankload.mat)*

Polyserial Correlation Matrix
1.000
0.041 1.000
0.524 –0.163 1.000
0.589 0.099 0.335 1.000
0.454 0.251 0.610 0.299 1.000
0.261 0.138 0.380 0.150 0.559 1.000
0.320 0.162 0.411 0.166 0.598 0.647 1.000

MODEL MODIFICATION

SEM software programs provide modification indices when the sample data do not fit the specified theoretical model. The modification index (MI) indicates how much the model-fit chi-square will be reduced if a path is added in the model. We generally select the MI with the largest value, add the path, and then re-run the analysis.

We should first look at the error covariance terms, because these are considered random sources of error, which might be related to observed variable relations. We can look next at the factor loadings of the indicator variables. Quite possibly, one or more of the selected indicator variables in the measurement models might not work when defining the latent variable. In addition, if we change the structural paths, then we are changing the core concept of the theoretical model.

The modification indices (MI) for the error covariance term between ADDRESS and AGE had the largest decrease in chi-square value (181.8). The next highest

MI was indicated for EMPLOY and ED (144.7), followed by EMPLOY and AGE (33.8). Because both of these are related to the *Ability* latent variable, we can add these and re-run the analysis. Command lines will need to be added to indicate the error covariances we want estimated between these observed variables. We will also lose a degree of freedom for each path added to the model. Therefore, our new degrees of freedom will be 13 − 3 = 10 degrees of freedom.

We felt that EMPLOY (years with current employer), ED (education level), ADDRESS (years at current address), and AGE were very much related to each other. We therefore added the following commands to the program to correlate their respective error covariance:

```
Let error covariance of EMPLOY and ED correlate
Let error covariance of ADDRESS and AGE correlate
Let error covariance of EMPLOY and AGE correlate
```

Our results continued to indicate a poor model fit (ML χ^2 = 47.75, df = 10, p > 0.00001, RMSEA = 0.072). Additional MI were indicated given the significant model-fit chi-square. The MI suggested to add error covariances between ADDRESS and EMPLOY (30.8), ADDRESS and ED (14.8), and ED and AGE (12.5). We therefore added three additional command lines to estimate their error covariance terms. The degrees of freedom drops to 10 − 3 = 7. These modifications also seemed reasonable given how years with current employer, years at current address, age, and education were related. We therefore added the following additional command lines to the SIMPLIS program:

```
Let error covariance of EMPLOY and ADDRESS correlate
Let error covariance of ADDRESS and ED correlate
Let error covariance of AGE and ED correlate
```

The final modified mixed variable model, which included all of these error covariance terms, indicated a good data to model fit (ML χ^2 = 5.59, df = 7, p = 0.587, RMSEA = 0.00). It is not unusual to add error covariances in a measurement model to obtain good data to model fit. Our extended effort in adding error covariance terms between observed variables in the model was due to the condition code indicating severe multicollinearity, but more specifically redundancy among the indicator variables. We desired to keep all of the observed variables in the model, but because they are related (correlated) it caused difficulty in obtaining parameter estimates. You could consider dropping a few of the indicator variables, which would simplify the model, but also affect the rationale for the theoretical model specification. Mixed variable models using raw data will use an asymptotic covariance matrix in addition to the sample covariance matrix, allow selection of a different estimation method, and compute a robust chi-square statistic. When possible, use the raw data set in LISREL9 to take advantage of these new features.

ROBUST CHI-SQUARE STATISTIC

Our SEM analysis required a polyserial correlation matrix because we had different types of observed variables (ordinal and continuous). We should therefore be reporting a robust chi-square statistic, not a normal theory chi-square statistic. How do you obtain the Satorra–Bentler chi-square robust statistic value? This option will vary depending upon your SEM software.

In LISREL, we first open the PRELIS system file, *bankloan.lsf*, and then save a covariance matrix (*bankloan.cov*) and an asymptotic covariance matrix (*bankloan.acm*) using the Statistics pull-down menu and *Output Option*. The SIMPLIS program is now modified to include the *Covariance matrix from file*, *Asymptotic Covariance Matrix from File*, and *Method of Estimation: Maximum Likelihood* commands. The computer output under *Goodness-of-Fit* statistics will now include the robust Satorra–Bentler scaled chi-square statistic. The updated SIMPLIS program with these commands would be:

```
Mixture Model using Polyserial Correlation Matrix
Observed Variables: AGE ED EMPLOY ADDRESS INCOME CREDDEBT OTHDEBT
Sample Size 723
Covariance matrix from file bankloan.cov
Asymptotic Covariance Matrix from File bankloan.acm
Method of Estimation: Maximum Likelihood
Latent Variables Ability Debt
Relationships
AGE ED EMPLOY ADDRESS INCOME = Ability
CREDDEBT OTHDEBT = Debt
Debt = Ability
Let error covariance of EMPLOY and ED correlate
Let error covariance of ADDRESS and AGE correlate
Let error covariance of EMPLOY and AGE correlate
Let error covariance of EMPLOY and ADDRESS correlate
Let error covariance of ADDRESS and ED correlate
Let error covariance of AGE and ED correlate
Number of Decimals = 3
Path Diagram
End of Problem
```

The final theoretical model indicated a good data to model fit with the Satorra–Bentler scaled chi-square statistic: $\chi^2 = 3.419$, $df = 7$, $p = 0.84$, which was less than the normal theory chi-square statistic: $\chi^2 = 5.69$, $df = 7$, $p = 0.57$. We should expect the robust statistic to indicate a better model fit because our data did not meet the assumptions of normal theory estimation given a mix of different variable types.

MIXTURE MODEL—LATENT CLASSES

Latent class analysis can be computed in several different software packages, for example, Mplus, SAS, and the R package (*poLCA*). Latent class analysis provides a method where *latent* groups are identified, rather than analyses with explicit defined group differences coded, for example, male vs female or experimental vs control (Dayton, 1999; Hagenaars & McCutcheon, 2002; McCutcheon, 1987). The observed dependent variables can be continuous, binary, ordinal, Poisson counts, or a combination of these variable types. It is used primarily when group membership is not known beforehand, that is, the variables determine group membership from the data's heterogeneity. Group membership is not observable but inferred, thus the term *latent class* (Lubke & Muthén, 2005). The latent class analyses of heterogeneous groups have also been referred to as mixture models (Muthén, 2002), and as a measurement model (Muthén & Muthén, 2006).

The latent class mixture model can be represented by a heterogeneous continuous variable distribution; however, the data suggest two different groups of individuals that are not directly observed. The two groups are termed latent classes ($c = 1$ and $c = 2$) with different means ($\mu 1$ and $\mu 2$). Dotted lines are used to indicate that these two latent class group distributions are not observed, while the solid line is a single distribution with a mixture of the two distributions (Figure 12.2). The latent class mixture model analysis can be used to determine the presence of latent classes and its associated parameter estimates. Assessment of the latent class distribution can be compared to the homogeneous distribution by a measure of data to model fit, where a c categorical latent variable is added with arrows indicating that the latent classes may differ (Clogg, 1995; Nylund, Asparouhov, & Muthén, 2007).

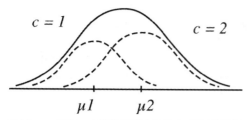

$c = 1$ $c = 2$

$\mu 1$ $\mu 2$

■ **Figure 12.2:** MIXTURE MODEL DISTRIBUTION

There are several issues related to conducting a latent class analysis. These are centered around the use of conditional probabilities, class membership determination, latent class size, number of latent classes specified, and graphing results. We will briefly address each of these issues next.

Conditional Probabilities

Latent class analysis is not the same as cluster analysis, factor analysis, or discriminant analysis. Latent class analysis is based on a statistical probability modeling approach. Cluster analysis is not based on a conditional probability statistical modeling approach; rather, it displays data clustered into groups around a centroid. Factor analysis techniques can be used with continuous or ordinal variables in SEM, but the analyses are focused on shared commonality of variables in defining a construct. Discriminant analysis is more similar to latent class analysis, in that observed variables are used to predict and classify group membership; however, the group membership is known and coded, so the emphasis is on actual versus predicted classification probabilities. Latent class analysis therefore extends this area of research using conditional probabilities across sets of variables for the number of classes selected. The probability of answering the first question yes rather than no is conditional on the next question also being answered yes, and so on. The pattern of conditional probability responses would be categorized differently depending upon the number of latent classes. Schumacker and Tomek (2012) and Schumacker (2013) cover the basic usage and formula for conditional probabilities in statistics.

Class Membership and Class Size

Mplus creates a data value that contains the original data, the probability of class membership, and a designation of the most probable class (modal class). The probability values across the number of latent classes will sum to 100%. If the distribution of probabilities of class membership makes sense, then the researcher must name the class membership. This subjective naming of the class is based on the questions or variables used in the latent class analysis. The face validity for naming the class is similar to the subjective naming of a construct in factor analysis. The probability of class membership into a modal class value provides for a summary of class size, whether a single class or two or more classes (Marcoulides & Heck, 2009; Schumacker & Pugh, 2013). In Figure 12.2, two classes are subsumed under a single homogeneous distribution. The latent class sizes do not have to be the same, for example, a "like math" latent class may have more students than a "do not like math" latent class based on answers to a survey questionnaire.

Number of Latent Classes

A key issue in latent class analysis is determining the number of classes in a distribution. A researcher should examine a class = 1 initially to determine whether any latent classes are subsumed in the heterogeneous distribution. Notice we are using the term heterogeneous in contrast to homogeneous. A homogeneous distribution would imply a single distribution, whereas a heterogeneous distribution implies the possibility of sub-groups. Mplus permits an easy way to change the number of classes and provides model-fit output to make a determination of the number of latent classes. The Vuong–Lo–Mendell–Rubin likelihood ratio test is used to determine the difference in loglikelihood chi-square for the models with a different number of latent classes specified. The associated bootstrap output tends to indicate a more robust result. So, when the two disagree, the bootstrap results are generally interpreted.

Graphing

The probabilities of each class across each variable in the model can be plotted. This provides an important visual display to show the class differences or similarities in the conditional probabilities. The Y axis represents the probability of an answer to a question for the latent classes, while the X axis indicates the question number. This also assumes we have decided on the right number of latent classes given the responses to the variables. The graph will visually show separation of the latent classes (groups) if present across the questions or variables.

Mplus Latent Class Example

Mplus provides numerous examples for using latent class analyses with continuous, binary, and count data (Muthén & Muthén, 2006). We will analyze a latent class example using the IBM SPSS data set, *bankloan.sav*. The data set contained variables related to default on bank loans. We were interested in the age, education, employment, years at address, income, debt to income ratio, debt to credit ratio, and default status (0 = no; 1 = yes) variables. The original data set contained $n = 850$ responses, but $n = 150$ were missing on the variable, *default*, so we used $n = 700$ cases in the latent class analysis.

We hypothesized that two distinct classes of people should be present in the data, Debt problems versus No Debt problems. We initially included all variables in the Mplus program, but failed to find two distinct classes. By a process of elimination, we dropped education, employment, address, othdebt, and default variables before achieving a clear latent class result indicating two distinct classes. We were surprised that education, employment, years at address, and default on loan did not improve the model fit.

The latent class model would be diagrammed as indicated in Figure 12.3. This is a continuous variable LCA similar to example 7.9 in the Mplus user guide (Muthén & Muthén, 2006), where many other types of LCA models are presented. The diagram shows the latent class, C, indicated by four continuous variables.

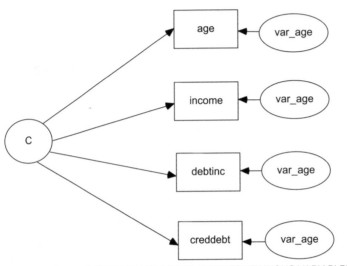

■ **Figure 12.3:** LATENT CLASS MODEL WITH CONTINUOUS VARIABLES

The final Mplus program does not require a model statement because automatic start values are used, rather than user-supplied values. Also, the variable means are given for each latent class, variances are set equal, and no correlation amongst variables within class is specified. The CLASSES = c(2) command specifies that we expect two latent classes to exist given the observed variables. The final Mplus program was written as follows:

```
TITLE: LCA with bankloan latent class indicators
DATA:FILE IS debt.dat;
VARIABLE:
NAMES ARE age,ed,employ,address,income,debtinc,creddebt,
     othdebt,default;
  USEVARIABLES ARE age income debtinc creddebt;
  CLASSES = c(2);
ANALYSIS:  TYPE = MIXTURE;
PLOT:     Type = plot3;
OUTPUT: TECH11 TECH14;
```

Results indicated a final class count and proportion as: Latent class 1, $n = 14$ (2%); Latent class 2, $n = 686$ (98%), with Entropy = .99 (indicates good separation of two latent class distributions). The model results indicated statistically significant continuous variables within each latent class. The computer output was as follows:

```
MPLUS OUTPUT

                                               Two-Tailed
                 Estimate     S.E.      Est./S.E.  P-Value
Latent Class 1
Means
   AGE            45.365     1.690       26.837     0.000
   INCOME        178.396    29.253        6.098     0.000
   DEBTINC        18.139     2.729        6.648     0.000
   CREDDEBT       11.811     1.196        9.875     0.000
Variances
   AGE            61.597     2.779       22.168     0.000
   INCOME        990.782   139.284        7.113     0.000
   DEBTINC        45.268     2.947       15.361     0.000
   CREDDEBT        2.313     0.231       10.007     0.000
Latent Class 2
Means
   AGE            34.644     0.301      115.088     0.000
   INCOME         42.871     1.088       39.396     0.000
   DEBTINC        10.099     0.255       39.587     0.000
   CREDDEBT        1.343     0.055       24.536     0.000
Variances
   AGE            61.597     2.779       22.168     0.000
   INCOME        990.782   139.284        7.113     0.000
   DEBTINC        45.268     2.947       15.361     0.000
   CREDDEBT        2.313     0.231       10.007     0.000
```

The statistical tests of model fit, that is, null hypothesis of single latent class versus alternative hypothesis of two latent classes, was $p < .02$. This suggests that we reject the null hypothesis (H_O) of a single class and accept the alternative hypothesis of two latent classes (H_A). The computer output for the Likelihood Ratio test for the null versus alternative hypothesis was shown as:

```
VUONG-LO-MENDELL-RUBIN LIKELIHOOD RATIO TEST FOR 1 (H0) VERSUS
   2 CLASSES
            H0 Loglikelihood Value                    -9820.235
            2 Times the Loglikelihood Difference        588.476
            Difference in the Number of Parameters           5
            Mean                                       -127.855
```

```
        Standard Deviation                        476.791
        P-Value                                    0.0178
LO-MENDELL-RUBIN ADJUSTED LRT TEST
        Value                                     571.043
        P-Value                                    0.0196
```

The parametric bootstrap computer results further supported the above statistical test.

```
PARAMETRIC BOOTSTRAPPED LIKELIHOOD RATIO TEST FOR 1 (H0) VERSUS
    2 CLASSES
            H0 Loglikelihood Value              -9820.235
            2 Times the Loglikelihood Difference   588.476
            Difference in the Number of Parameters       5
            Approximate P-Value                    0.0000
            Successful Bootstrap Draws                   5
```

This latent class analysis points out a few key issues when conducting the analysis. First, we had to handle missing data, otherwise it interferes with the classification process. Second, the variables we originally thought would provide two latent classes did not do so. So, by a process of elimination we arrived at the observed variables that showed two latent classes. We then followed through by naming these two latent classes, as originally hypothesized. Finally, although two latent classes emerged, the latent class count was $n = 14$ (2%) versus $n = 686$ (98%). So, out of the 700 individuals, only 2% would be considered as having a debt problem, while 98% would be considered as having no debt problems. The raw data set indicated that $n = 517$ (74%) had no defaults, while $n = 183$ (26%) had defaulted on a loan. We expected better latent class results given this knowledge. This also points out the importance of model specification, that is, selection of the variables for making the latent class distinction.

SUMMARY

The SEM mixture model permits continuous and categorical variables to be used in a theoretical model. The mixture model, however, uses a different correlation matrix than the traditional Pearson correlation matrix with means and standard deviations. Consequently, you will need to read in a raw data set and output a polyserial correlation matrix. Additionally, you will need to save a covariance matrix and an asymptotic covariance matrix to include in your SEM program, along with the appropriate estimation method to obtain the Satorra–Bentler scaled chi-square statistic for interpretation of the mixture model.

The mixture model using different types of variables illustrates many important steps in SEM modeling. First, we need to create the correct type of variance–covariance matrix for the mixture model. Second, we need to address missing data and multicollinearity amongst the variables. In our example, observed variable redundancy was an issue, most likely due to linear dependency. We used excessive MI corrections amongst the error covariances to adjust somewhat for the variable redundancy. The MI corrections require justification because they go beyond random error, and compensate for variable redundancy. The estimation method selected also impacted computing unbiased parameter estimates. Finally, a robust chi-square test of model fit was important given the type of matrix used.

Latent class analysis requires picking the right variables for the model, determining the correct number of latent classes, and selecting names for the latent class. The model specification is therefore an important first step in latent class analysis. Latent class analysis also requires consideration of whether you are using binary, ordinal, or continuous variables. The Mplus software permits running the program with different sets of variables, and more importantly, specifying a different number of latent classes. Computation time increases dramatically as you increase the number of classes and the number of variables.

From a modeling perspective, you also need to consider missing data, start values, number of iterations, number of latent classes, and the size of each latent class. When using continuous variables, a decision has to be made on whether the variables are correlated within the latent classes. Overall, many different types of models are possible, so attention to detail is required.

EXERCISES

1. Mixed Variable Model—Different Variable Types

Given the mixed variable model in Figure 12.4, write an SEM program to test the mixed variable model. Robust statistics require the raw data file, so **no** Satorra–Bentler scaled chi-square is computed. The mixed variable model has six observed variables (*Age, Gender, Degree, Region, Hours*, and *Income*) that define two latent variables (*Person* and *Earning*). A polyserial correlation matrix was created where CO indicates a continuous variable and OR indicates a categorical variable. *Age* (CO), *Gender* (OR), and *Degree* (OR) define Personal Characteristics, an independent latent variable (*Person*). *Region* (OR), *Hours* (CO), and *Income* (CO) define the dependent latent variable Earning Power (*Earning*). Personal Characteristics (*Person*) is hypothesized to predict Earning Power (*Earning*).

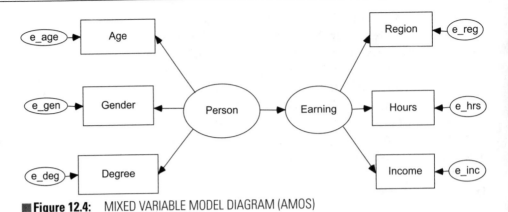

■ **Figure 12.4:** MIXED VARIABLE MODEL DIAGRAM (AMOS)

The information needed for the SEM mixed variable model program is:

```
Observed Variables: Age Gender Degree Region Hours Income
Correlation Matrix
 1.000
 0.487 1.000
 0.236 0.206 1.000
 0.242 0.179 0.253 1.000
 0.163 0.090 0.125 0.481 1.000
 0.064 0.040 0.025 0.106 0.136 1.000
Means 15.00 10.000 10.000 10.000 7.000 10.000
Standard Deviations 10.615 10.000 8.000 10.000 15.701 10.000
Sample Size 600
```

2. Mixture Model—Binary Latent Class

The binary latent class is modeled similar to the continuous variable example in this chapter, except the *Categorical* command is used to specify the variables as binary or ordinal (see example 7.3 in the Mplus user's guide). We used the data set from www.math.smith.edu/r/data/help.csv to conduct an LCA using the following 4 variables: homeless, cesdcut, satreat, and linkstat, with $n = 431$ data points ($n = 453$ with 22 missing cases). The binary variables are coded as follows: homeless (1 = yes; 0 = no); cesdcut (1 = high > 20; 0 = low < 20); satreat (1 = substance abuse treatment; 0 = no treatment); linkstat (1 = linked primary care; 0 = no primary care). For convenience, the reduced data set, *newds*, is provided on the book website for this exercise. Descriptive statistics indicated 47% homeless; 82% high cesdcut scores; 29% substance abuse treatment; and 38% linked to primary care. We hypothesized that there exist two latent classes: homeless with substance abuse and homeless without substance abuse. Do the results support two latent classes? The MPlus program was written as follows:

```
TITLE:LCA with binary latent class indicators
DATA:FILE IS newds.dat;
VARIABLE:   NAMES ARE homeless cesdcut satreat linkstat;
    CLASSES = c (2);
    CATEGORICAL = homeless cesdcut satreat linkstat;
ANALYSIS:   TYPE = MIXTURE;
OUTPUT:     TECH11 TECH14;
```

SUGGESTED READINGS

Bagley, M. N., & Mokhtarian, P. L. (2002). The impact of residential neighborhood type on travel behavior: A structural equations modeling approach. *The Annals of Regional Science, 36,* 279–297.

Loken, E. (2004). Using latent class analysis to model temperament types. *Multivariate Behavioral Research,* 39(4), 625–652.

Rindskopf, D., & Rindskopf, W. (1986). The value of latent class analysis in medical diagnosis. *Statistics in Medicine,* 5, 21–27.

Uebersax, J. S., & Grove, W. M. (1990). Latent class analysis of diagnostic agreement. *Statistics in Medicine,* 9, 559–572.

REFERENCES

Clogg, C. C. (1995). Latent class models. In G. Arminger, C. C. Clogg, & M. E. Sobel (Eds.), *Handbook of statistical modeling for the social and behavioral science* (pp. 311–359). New York: Plenum.

Dayton, M. C. (1999). *Latent class scaling analysis.* Series: Quantitative Applications in the Social Sciences. Thousand Oaks, CA: Sage Publications.

Hagenaars, J. A., & McCutcheon, A. L. (2002). *Applied latent class analysis.* Cambridge: Cambridge University Press.

Jöreskog, K., & Sörbom, D. (1996). *PRELIS2: User's reference guide.* Chicago, IL: Scientific Software International.

Jöreskog, K., Sörbom, D., du Toit, S., & du Toit, M. (2001). *LISREL8: New statistical features.* Chicago, IL: Scientific Software International.

Lubke, G. H., & Muthén, B. (2005). Investigating population heterogeneity with factor mixture models. *Psychological Methods,* 10, 21–39.

Marcoulides, G. A., & Heck, R. H. (2009). Educational applications of latent growth mixture models. In T. Teo, & M. S. Khine (Eds.), *Structural equation modeling in educational research: Concepts and applications* (pp. 346–366). Rotterdam, NL: Sense Publishers.

McCutcheon, A. L. (1987). *Latent class analysis*. Series: Quantitative Applications in the Social Sciences. Thousand Oaks, CA: Sage Publications.

Muthén, B. O. (2002). Latent variable mixture modeling. In G. A. Marcoulides, & R. E. Schumacker (Eds.), *New developments and techniques in structural equation modeling* (pp. 1–33). Mahwah, NJ: Lawrence Erlbaum Associates.

Muthén, L. K., & Muthén, B. O. (2006). *Mplus user's guide* (4th edn.). Los Angeles: Muthén & Muthén.

Nylund, K. L., Asparouhov, T., & Muthén, B. O. (2007). Deciding on the number of classes in mixture modeling: A Monte Carlo simulation. *Structural Equation Modeling*, 14(4), 535–569.

Schumacker, R. E. (2013). *Learning statistics using R*. Thousand Oaks, CA: Sage Publications.

Schumacker, R. E., & Pugh, J. (2013). Identifying reading and math performance in school systems with latent class longitudinal growth modeling. *Journal of Educational Research and Policy Analysis* 13(3), 51–62.

Schumacker, R., & Tomek, S. (2012). *Understanding statistics using R*. New York: Springer-Verlag.

Chapter 13

MULTI-LEVEL MODELS

CHAPTER CONCEPTS

It's just regression
Multi-level model example–observed variables
  Null model (baseline)
  Intraclass correlation
  Deviance statistic
Multi-level model example–latent variables

IT'S JUST REGRESSION

Multi-level models are considered a type of general linear model when using observed variables, which are named because of the hierarchical nature of data in a nested research design (Bickel, 2007). For example, in education a student's academic achievement is based on instruction in a classroom, so students are nested in classrooms, teachers are nested within schools, and schools are nested within districts. The nested research design is in contrast to a crossed research design where every level is represented. In multi-level models our interest is in the effects at different levels given the stratification or clustered nature of the data. A simple schematic will illustrate multi-level versus crossed designs.

Multi-level Design

Four teachers work at two schools; however, teachers 1 and 2 are in School A, while teachers 3 and 4 are in School B.

School	A		B	
Teacher	1	2	3	4

Crossed Design

Four teachers work at two schools; however, all four teachers are in both schools.

School	A	B
Teacher	1 2 3 4	1 2 3 4

Multi-level modeling involves predicting variance at different levels determined by the stratification or clusters. In the basic two-level model, a test of significant variation within groups is tested. There may be significant intercept variation (γ_{00}) or between-group slope variation (γ_{11}). If intercept and slope variation is not significant by group, then there is little reason to use multi-level modeling; instead, an OLS regression analysis is sufficient. Consequently, group-level properties of the response variable (dependent) should involve checking the intraclass correlation coefficient, that is, how much variance in the response variable is explained by group membership. The higher the intraclass correlation coefficient, the more score variance is attributed to the stratification or cluster (grouping variable). A rule of thumb has been that if the intraclass correlation is greater than 10%, a multi-level analysis should be conducted, rather than multiple regression (Bickel, 2007). In analysis of variance (ANOVA) terms, we would express the intraclass correlation coefficient ($\hat{\rho}_{ICC1}$) using the mean square between (MS_B), mean square within (MS_w), and the degrees of freedom between groups (df_B), as

$$\hat{\rho}_{ICC1} = \frac{MS_B - MS_W}{MS_B + (df_B)MS_W}$$

A *design effect* is also useful in understanding the difference between a multi-level nested design with stratification or clusters compared to a simple random sample. In a simple random sample, equal weighting of sample size would occur when computing parameter estimates; however, this could lead to biased parameter estimates if the random sampling over-sampled certain subpopulations. For example, more males are sampled than females when surveying opinions on voting rights. If the intraclass correlation coefficient is greater than 10%, then scores within a stratification or cluster are more similar than scores selected from a simple random sample. This indicates that the higher the intraclass correlation coefficient, the more scores are related to a cluster. A simple calculation using the intraclass correlation coefficient and sample size of the cluster reveals whether scores are more related to a cluster, or independent of a cluster:

$$\text{DEFF} = \hat{\rho}_{ICC1}(n_c - 1) + 1$$

If the intraclass correlation is zero, then DEFF = 1, indicating that scores are independent of clusters. For DEFF > 1, the ratio of the variance in a multi-level model with clusters compared to that of a simple random sample can be interpreted. For example, given an intraclass correlation of .10, and 25 cases per cluster, the DEFF is

DEFF = .10(25 − 1) + 1 = 3.4

This indicates that the variance in a multi-level model is 3 times greater than in the simple random sample (n = 250). The *effective sample size*, which takes into account the score dependence and cluster size, is computed by dividing the simple random sample size by the DEFF value: 250 / 3.4 = 73.5 ~ 74. Power calculations are usually based on this effective sample size (n = 74), not the original sample size (n = 250). The sampling error in the multi-level model is comparable to that expected in a simple random sample of 74 subjects. Some software programs that calculate power and effect size, permit calculations based on number of stratifications or clusters in the multi-level model.

Next, one should determine the reliability of the group means on the response variable. The reliability coefficient should be equal to or greater than .70, which would indicate that group means (intercept values) can be consistently differentiated (Bliese, 1998). Shrout and Fleiss (1979) provided several different intraclass correlations; however, the one of interest in multi-level modeling is τ_{ICC2} / ($\tau_{00} + \sigma^2$). A second intraclass correlation coefficient ($\hat{\rho}_{ICC2}$), using MS_B and MS_W, indicates whether groups can be reliably differentiated on the response variable (Bliese, 2000):

$$\hat{\rho}_{ICC2} = \frac{MS_B - MS_W}{MS_B}$$

Finally, the variance of the intercept values for the groups should be statistically significant (greater than zero). Obviously, if the intercept or group means are not significantly different, then a common intercept term can be used in an OLS regression equation.

The intercept-only model (null model) does not contain any predictor variables, thus assesses the random intercept variance for the groups. This model estimates the variability in the mean Y values related to the total variability. $Y_{ij} = \gamma_{00} + u_{0j} + r_{ij}$, where subscripts indicate individual (i) and group (j) designations. The equation shows that the response scores are a function of a common intercept term and two error terms: the between-group error (u_{0j}) and the within-group error term (r_{ij}). The null model therefore indicates the intercept variance of each group around the overall common intercept (τ_{00} associated with u_{0j}), and the individual score variance from the group mean (σ^2 associated with r_{ij}).

MULTI-LEVEL MODEL EXAMPLE—OBSERVED VARIABLES

The multi-level *null model* is a preliminary first step in a multi-level analysis because it provides important information about the variability of the dependent variable. You should always create a null model (intercept-only) to serve as a baseline for comparing additional multi-level models when you add variables to test whether they significantly reduce the unexplained variance in the dependent variable (response or outcome variable).

Null Model (Baseline)

The multi-level model is based on the mouse data set in LISREL (Jöreskog & Sörbom, 1993), which is a nested data set with 9 weight measurements taken at 9 time periods on 82 mice. The data set should contain $n = 738$ rows of data (9 time periods × 82); however, the data set only contains $n = 698$ rows of data because some weights are missing for the mice. The variables in the multi-level model are id, weight, time, and gender, which are saved in the file, *mouse.txt*, on the book website. The multi-level model is appropriate because weight is nested within time.

The multi-level model was run using LISREL–PRELIS, and the commands saved in the file, *PRELIS.txt*, on the book website. The null (baseline) model only includes the intercept term. The computer output should show the fixed and random results for the null (baseline) model with the intercept-only (β_0), and the two random variance terms, $u_{ij} + e_{ij}$. The null model equation is $Y_{ij} = \beta_0 + u_{ij} + e_{ij}$. The equation results are $Y_{ij} = 28.63 + 11.32 + 130.32$, with deviance (–2LL) = 5425.49. Our goal is to reduce the deviance (–2LL) value by adding additional variables.

The second multi-level analysis includes adding *time* to the multi-level model equation. The multi-level model equation is $Y_{ij} = \beta_0 + \beta_1 \ Time_{ij} + u_{ij} + e_{ij}$. This equation assesses the weight gain (or loss) over the nine measurement time periods. The equation results are $Y_{ij} = 9.09 + 4.09 + 20.69 + 16.46$, with deviance (–2LL) = 4137.57. We are now able to assess the statistical significance of adding a variable, *time*, to the equation. A loglikelihood chi-square difference test is used to test the statistically significant difference, $\Delta\chi^2 = \chi^2_0 - \chi^2_1$, with $df = 1$. $\Delta\chi^2 = 5425.49 - 4137.57 = 1287.92$. The –2LL chi-square difference of 1287.92 is greater than the critical chi-square of 3.84, $df = 1$, at the .05 level of significance.

We can therefore conclude that adding the variable *time* significantly reduced the amount of unexplained variance.

We run the multi-level model equation again, but now we add *gender* to the equation. The multi-level model equation is now $Y_{ij} = \beta_0 + \beta_1 Time_{ij} + \beta_2 Gender_{ij} + u_{ij} + e_{ij}$. The equation results are $Y_{ij} = 9.07 + 4.08 + 1.42 + 18.68 + 16.46$, with deviance (–2LL) = 4129.94. We can once again conduct a –2LL chi-square difference test to determine if adding the variable *gender* decreases the amount of unexplained variance. $\Delta\chi^2 = 4137.57 - 4129.94 = 7.63$. The –2LL chi-square difference of 7.63 is greater than the critical chi-square of 3.84, $df = 1$, at the .05 level of significance. Gender did provide an additional reduction in the unexplained variance of the dependent variable.

We could have continued adding observed variables to the multi-level modeling equation, but stopped to consolidate the results. Obviously, if you added additional hypothesized variables, then you would expect to obtain a better prediction of the unexplained variance in the response variable (*weight*). The three different multi-level analysis results for an intercept-only model (model 1), intercept and time model (model 2), and an intercept, time, and gender model (model 3) are summarized in Table 13.1. Model 1 provided a baseline model in which to determine if additional variables help in reducing the amount of variance in *weight*. Model 2 with *time* added substantially reduced the unexplained variance in *weight* ($\chi^2 = 1287.92$, $df = 1$). Model 3 with *gender* added also significantly reduced the amount of unexplained variance in *weight* ($\chi^2 = 7.63$, $df = 1$). Therefore, mouse *weight* variability is statistically significantly explained by *time* and *gender* observed variables.

Table 13.1: Multi-level Models for Mouse Weight

Multi-level Model	Model 1	Model 2	Model 3
	Intercept Only	Intercept + Time	Intercept + Time + Gender
Intercept Only (B_0)	28.63 (.57)	9.09 (.60)	9.07 (.58)
Time (B_1)		4.09 (.06)	4.08 (.06)
Gender (B_2)			1.42 (.50)
Level 2 error variance (e_{ij})	130.32	16.46	16.46
Level 3 error variance (u_{ij})	11.33	20.69	18.68
ICC	.079 (8%)	.556 (56%)	.532 (53%)
Deviance (–2LL)	5425.49	4137.57	4129.94
df	3	4	5
Chi-square Difference ($df = 1$)		1287.92	7.63

Note: $\chi^2 = 3.84$, $df = 1$, $p = .05$.

Intraclass Correlation

The presence of a significant intraclass correlation coefficient (ICC) indicates the need to employ multi-level modeling rather than OLS regression. The main difference is in the standard errors of the parameters. Multiple regression has smaller standard error estimates when the intraclass correlation coefficient is statistically significant, which would inflate (bias) the regression weights. The intraclass correlation coefficient, using our results, is computed in SEM as

$$ICC = \frac{\Phi_3}{\Phi_3 + \Phi_2} = \frac{Tau - Hat(Level - 3)}{Tau - Hat(Level - 3) + Tau - Hat(Level - 2)}$$
$$= \frac{11.33}{11.33 + 130.32} = .079$$

Therefore, 8% of the variance in *weight* is explained in the baseline model. It jumps dramatically when adding *time* as an explanatory variable to 56% variance in *weight*, explained as a function of *time*. It drops modestly to 53% when adding *gender* to the equation. The 3% difference is not enough to infer a non-significant effect; therefore, *time* and *gender* significantly explain 53% of the variance in mouse *weight*.

The intraclass correlation coefficient (ICC) measures the relative homogeneity within groups in ratio to the total variation. If your SEM software does not compute the ICC, SPSS has a drop-down menu option for computing the intraclass correlation coefficient in your data. You can also compute the ICC by hand calculations given the formula above using the tau estimates provided by the SEM program.

If the intraclass correlation coefficient is large and positive, then there is no variation within the groups, but group means are different. The ICC will have a large negative value when group means are the same, but there is greater variation within groups. The ICC maximum value is 1.0, but its maximum negative value is $(-1/(n - 1))$. A negative intraclass correlation coefficient occurs when the between-group variation is less than the within-group variation. This indicates that a non-random third variable is present which suppresses group differences.

Deviance Statistic

The deviance statistic is computed as $-2LL$, which is used to test for statistical difference in models between Model 1 (intercept), Model 2 (intercept + time), and Model 3 (intercept + time + gender). The chi-square difference ($\Delta\chi^2$) is usually tested against the chi-square value of 3.84, *df* = 1, at the *p* = .05 level of significance to test whether added variables in the equation reduced the amount of unexplained variance in mice *weight*. The baseline deviance value was 5425.49. The chi-square difference test between this baseline deviance statistic and the second equation deviance value with time (–2LL = 4137.57) indicated a difference of 1287.92, which is statistically significantly different than the tabled critical chi-square value of 3.84. Consequently, *time* was a significant predictor variable of mice *weight*. The model with *time* and *gender* indicated a deviance statistic of 4129.94 and had a difference from the previous deviance statistic of 7.63, which was also statistically significantly different from the critical tabled chi-square of 3.84. Consequently, *time* and *gender* were statistically significant predictor variables of mice *weight*.

MULTI-LEVEL MODEL EXAMPLE – LATENT VARIABLES

Our example will be a single-factor model, *math ability*, with three indicator variables: math1, math2, and math3. The strata or clusters are indicated by 49 different schools. We are hypothesizing that the factor loadings and factor variances between groups and within groups are equal. The confirmatory factor model is diagrammed in Figure 13.1.

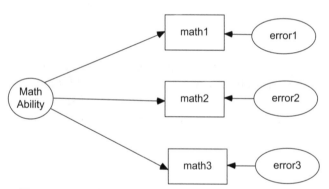

■ **Figure 13.1:** CONFIRMATORY FACTOR MODEL (AMOS)

The SIMPLIS program will model the between and within variance to obtain the two error variances. The matrices are obtained by the following: $\Sigma_B = \lambda \Psi \lambda' + D_B$ and $\Sigma_W = \lambda \Psi \lambda' + D_W$, where D_B and D_W represent the two unique diagonal error variance matrices. The SIMPLIS program to conduct the two-level (between and within) multi-level model for a single factor was saved as *multilevel.txt* on the book website. The SIMPLIS program is given here:

```
Multi-level Model - Single Factor with continuous variables -
school differences
Group 1: Between Schools (Level 2)
Observed variables; math1 math2 math3
Sample size: 24
Covariance matrix
3.38885
2.29824 5.19791
2.31881 3.00273 4.69663
Latent variables: ability
Relationships
math1 = 1*ability
math2-math3 = ability
Group 2: Within Schools (Level 1)
Sample size: 1192
Covariance matrix
47.04658
38.56798 55.37006
30.81049 36.04099 40.71862
Set the variance of ability free
Set the error variance of math1 free
Set the error variance of math2 free
Set the error variance of math3 free
Path diagram
End of Problem
```

The results are reported in Table 13.2 for the factor loadings, standard errors, and the two unique error variance matrix values. The factor loadings and factor variance were different across the schools, as indicated by the statistically significant z values. Findings therefore indicate that *math ability* varied across the 49 different schools.

SUMMARY

A researcher should understand that a multi-level model is useful when a nested research design strategy is employed. Otherwise, a general linear model will suffice. Multi-level models are popular today, but a researcher should first check the ICC to determine if a multi-level analysis is necessary. There are several HLM software programs that use observed variables in a nested research design (HLM, MLwiN, etc.), which have become increasingly popular in repeated measures, survey, and education data analysis because of the hierarchical research design.

Table 13.2: Multi-level SEM—Single-Factor Model

Factor Loadings	Estimate	Standard Error	Standardized	z	p
$\lambda 11$	1	—	.84		
$\lambda 21$	1.17	.20	.90	5.77	< .0001
$\lambda 31$	.937	.17	.84	5.51	< .0001
Factor Variance					
ψ	32.90	13.11		2.51	< .05
Error Variance (between)					
math1	1.41	.67		2.10	.035
math2	2.175	.96		2.25	.03
math3	2.15	.80		2.67	.007
Error Variance (within)					
math1	14.09	5.79		2.43	.02
math2	10.26	6.27		1.64	.10
math3	11.91	4.99		2.39	.02
ML Chi-square	1.01				
df	2				
p	.99				

In SEM, we also refer to this type of model as a multi-level model; however, we can conduct multi-level modeling with observed and/or latent variables. In the research literature this type of model is referred to by many different names, for example, hierarchical linear, random-coefficient, variance-component modeling, or HLM. SEM software provides additional features for multi-level modeling that involve using categorical variables, covariates, factor models, and structural models (Muthén & Muthén, 2006).

Several textbooks introduce and present excellent multi-level examples, so we refer you to those for more information on the analysis of multi-level models in SEM (Heck & Thomas, 2000; Hox, 2002). SEM software programs provide examples on multi-level modeling that include an overview of multi-level modeling; differences between OLS and multi-level random coefficient models (MRCM); multi-level latent growth curve models; testing of contrasts; analysis of two-level repeated measures data; multivariate analysis of educational data; multi-level models for categorical response variables; and examples using air traffic control data, school, and survey data. We have provided a few journal article references that have used the multi-level approach to show different applications across academic disciplines.

EXERCISES

1. Multi-level Model—Observed Variables

The data set for the multi-level model analysis is from LISREL9 in the *mlevelex* folder, *income.psf*, which is saved as *income.data*. The data set contains the variables *region*, *state*, *age*, *gender*, *marital*, etc. There are 9 regions with 51 states nested within the regions. The sample size is $n = 6062$. It is hypothesized that income varies by state within region.

Run your SEM program as follows:

a. Model 1 as an intercept-only model with *income* as the response variable; *state* and *region* as levels 2 and 3, respectively.
b. Model 2 will add *gender* as a fixed variable.
c. Model 3 will add *marital* as a fixed variable.

Summarize the results from the Model 1, Model 2, and Model 3 SEM programs in a table.

2. Multi-level Model—Latent Variable

This exercise is modeled after example 9.6 in the Mplus user's guide (Muthén & Muthén, 2006), which involves a two-level (within and between) factor model with continuous indicator variables, a random intercept, and covariate variables in an SEM multi-level latent variable modeling. It is also similar to an example described in Heck (2001) on leadership. Figure 13.2 displays the two-level model structure. The Mplus program is listed below.

a. Are the intraclass correlation coefficients > .10 so we can conduct a multi-level modeling of the variance across the 10 groups?
b. What is the chi-square model-fit value? Is the chi-square test of model fit non-significant, which indicates a good data to model fit?
c. Are the X1 and X2 within-model covariates statistically significant?
d. Are the Y1–Y4 factor loadings significantly different in the between factor model?
e. Are the Y1–Y4 intercepts significantly different in the between-factor model?

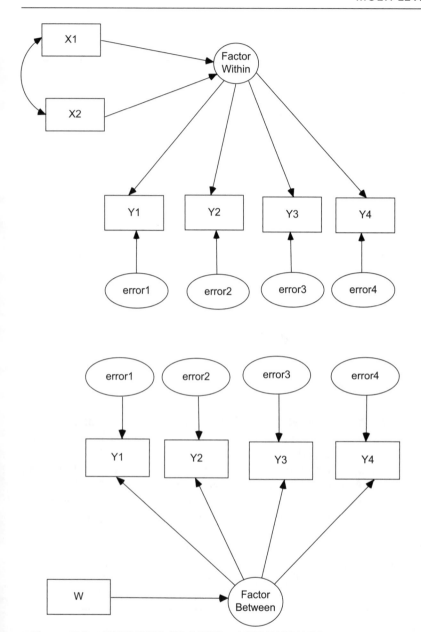

■ **Figure 13.2:** MULTI-LEVEL CFA MODEL—LATENT VARIABLE

Mplus Program

```
TITLE:Two -level CFA with continuous indicators, random inter-
   cept, and covariates
DATA:FILE IS ex9.6.dat;
VARIABLE:NAMES ARE y1-y4 x1 x2 w clus;
Usevariables are ALL;
WITHIN = x1 x2;
BETWEEN = w;
CLUSTER = clus;
ANALYSIS:TYPE = TWOLEVEL;
MODEL:
%WITHIN%
fw BY y1-y4;
fw ON x1 x2;
%BETWEEN%
fb BY y1-y4;
y1-y4@0;
fb ON w;
OUTPUT: SAMPSTAT STANDARDIZED;
```

SUGGESTED READINGS

Bryan, A., Schmiege, S. J., & Broaddus, M. R. (2007). Mediational analysis in HIV/AIDS research: Estimating multivariate path analytic models in a structural equation modeling framework. *AIDS Behavior*, 11, 365–383.

Everson, H. T., & Millsap, R. E. (2004). Beyond individual differences: Exploring school effects on SAT scores. *Educational Psychologist*, 39(3), 157–172.

Kidwell, R. E. Jr., Mossholder, K. W., & Bennett, N. (1997). Cohesiveness and organizational citizenship behavior: A multilevel analysis using work groups and individuals. *Journal of Management*, 23(6), 775–793.

Trautwein, U., Ludtke, O., Schnyder, I., & Niggli, A. (2006). Predicting homework effect: Support for a domain-specific, multilevel homework model. *Journal of Educational Psychology*, 98, 438–456.

REFERENCES

Bickel, R. B. (2007). *Multilevel analysis for applied research: It's just regression.* New York: Guilford.

Bliese, P. D. (1998). Group size, ICC values, and group-level correlations: A simulation. *Organizational Research Methods*, 1, 355–373.

Bliese, P. D. (2000). Within-group agreement, non-independence, and reliability: Implications for data aggregation and Analysis. In K. J. Klein, & S. W. Kozlowski (Eds.), *Multilevel theory, research, and methods in organizations* (pp. 349–381). San Francisco, CA: Jossey-Bass.

Heck, R. H. (2001). Multilevel modeling with SEM. In G. A. Marcoulides, & R. E. Schumacker (Eds.), *New developments and techniques in structural equation modeling*. Mahwah, NJ: Lawrence Erlbaum Associates.

Heck, R. H., & Thomas, S. L. (2000). *An introduction to multilevel modeling techniques*. Mahwah, NJ: Lawrence Erlbaum.

Hox, J. (2002). *Multilevel analysis: Techniques and applications*. Mahwah, NJ: Lawrence Erlbaum.

Jöreskog, K. G., & Sörbom, D. (1993). *LISREL 8: Structural equation modeling with the SIMPLIS command language*. Chicago: Scientific Software International.

Muthén, L. K., & Muthén, B. O. (2006). *Mplus user's guide* (4th edn.). Los Angeles: Muthén & Muthén.

Shrout, P. E., & Fleiss, J. L. (1979). Intraclass correlations: Uses in assessing rater reliability. *Psychological Bulletin*, 86, 420–428.

Chapter 14

LATENT GROWTH MODELS

CHAPTER CONCEPTS

Repeated measures designs
Latent growth curve model–intercept and slope
  Model modification
Latent growth curve model–group differences

REPEATED MEASURES DESIGNS

Repeated measures analysis of variance has been widely used with observed variables to statistically test for changes over time. Lomax and Hahs-Vaughn (2012) explain both the one-factor and mixed repeated measures designs. The one-factor or subjects only repeated measures across time is written in terms of $Y_{ij} = \mu + \alpha_j + s_i + (s\alpha)_{ij} + \varepsilon_{ij}$. The equation signifies that the Y scores for each individual are based on the overall population mean (μ), a fixed effect for each time period (α), a random effect for each subject (s), an interaction between subject and time period ($s\alpha$), and a random residual error (ε).

The mixed repeated measures design includes a between group factor, as well as the within repeated measures factor. The mixed repeated measures design equation is written in terms of $Y_{ij} = \mu + \alpha_j + s_{i(j)} + \beta_k + (\alpha\beta)_{jk} + (\beta s_{ki_{(i)}}) + \varepsilon_{ijk}$. The mixed design equation signifies that the Y scores for each individual are based on the overall population mean (μ), a fixed effect for each time period (α), a subject effect nested within each group ($s_{i(j)}$), group effect (β), interaction of time and group ($\alpha\beta$), interaction of within subject repeated measures and between subject ($\beta s_{ki(j)}$), and a random residual error for each individual time and group cell. The mixed repeated measures design assesses individual growth or change across time, as well as assesses group membership change across time, for example, boys versus girls.

In the multi-level modeling chapter, time was treated as a nested effect. In the mixed repeated measures design, subjects are considered nested in groups across time, which permits testing group differences. The advantage of repeated measures designs is that the subjects are their own control, thus sample size requirements are generally lower than for a simple random sample. However, when testing mixed models of subject and group across time, it is wise to increase the sample size due to testing interaction effects. The assumptions of independent observations, homogeneity of variance, normality, and sphericity apply to repeated measures designs using observed variables and group effects, and carry over to SEM applications using observed variables and latent variables. SEM, unlike ANOVA, doesn't rely on the correct ratio of means squares (MS), and the F-test for statistical significance.

SEM advances the longitudinal analysis of data to include latent variable growth over time while modeling both individual and group changes using slopes and intercepts (Byrne & Crombie, 2003; Duncan, Duncan, & Strycker, 2006; McArdle & Epstein, 1987; Stoolmiller, 1995). Latent growth curve analysis conceptually involves two different analyses. The first analysis is the repeated measures of each individual across time that is hypothesized to be linear or non-linear. The second analysis involves using the individual parameters (slope and intercept values) to determine the difference in growth from a baseline measure. The latent growth curve model (LGM) represents differences over time that take into account means (intercepts) and rate of change (slopes) at the individual or group level (Schumacker, 2016).

The LGM permits an analysis of individual parameter differences, which is critical to any analysis of change. It describes not only an individual's growth over time (linear or non-linear), but also detects differences in individual parameters over time. LGM using structural equation modeling can test the type of individual growth curve, use time varying covariates, establish the type of group curve, and include interaction effects in latent growth curves (Li, Duncan, Duncan, Acock, Yang-Wallentin, & Hops, 2001). The LGM approach, however, generally requires large samples, multivariate normal data, uses equal time intervals for all subjects, and involves change that occurs as a result of the time continuum (Duncan & Duncan, 1995).

LATENT GROWTH CURVE MODEL—INTERCEPT AND SLOPE

The latent growth curve model illustrates the use of slope and intercept as latent variables to model differences over time. The data set contains 168 adolescent responses over a 5-year period (age 11 to age 15) regarding the tolerance toward deviant behaviors, with higher scores indicating more tolerance of such behavior. The data were transformed (i.e., log X) to create equal interval linear measures from ordinal data. The slope parameters are coded 0, 1, 2, 3, and 4 to establish a linear trend with zero used as a common starting point. Other polynomial coefficients could be used to test for a quadratic or cubic trend. The intercept parameters are coded 1, 1, 1, 1, and 1 to indicate means for the different age groups. The latent growth curve model is diagrammed in Figure 14.1.

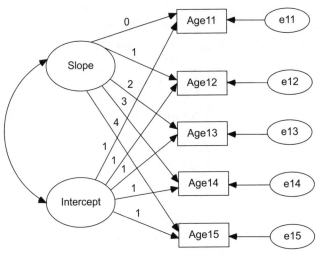

■ **Figure 14.1:** LATENT GROWTH CURVE MODEL (AMOS)

A SIMPLIS program was created to show how these parameters are stipulated for the two latent variables, *slope* and *intercept*. It also includes a command to correlate slope and intercept (curved arrow in diagram) and a special term, CONST, to designate means. The SIMPLIS latent growth curve model program is available on the book website, and was written as:

```
Latent Growth Model
Observed Variables: age11 age12 age13 age14 age15
Sample size 168
Correlation matrix
1.000
 .161 1.000
 .408 .348 1.000
 .373 .269 .411 1.000
 .254 .143 .276 .705 1.000
Means .201 .226 .326 .417 .446
Standard deviations .178 .199 .269 .293 .296
Latent Variables: slope intercept
Relationships:
```

```
age11 = CONST + 0 * slope + 1 * intercept
age12 = CONST + 1 * slope + 1 * intercept
age13 = CONST + 2 * slope + 1 * intercept
age14 = CONST + 3 * slope + 1 * intercept
age15 = CONST + 4 * slope + 1 * intercept
Let slope and intercept correlate
Path Diagram
End of Problem
```

The SIMPLIS results indicated a poor model fit (chi-square = 49.74, $df = 7$, $p = 0.00$). The correlation between the intercept values (group means) and the slope (linear growth) was zero, indicating that level of tolerance at age 11 did not predict growth in tolerance across the other age groups. However, the group means indicated otherwise, so model modification was conducted; the means for each age were shown on the computer output as:

age11	age12	age13	age14	age15
0.20	0.23	0.33	0.42	0.45

Modification indices were indicated that recommended correlating the error covariance between age 11 and age 12, as well as between age 14 and age 15. It is not uncommon to correlate lagged error terms in repeated measures designs. These time periods are apparently the two transition periods in the latent growth curve model where more measurement disturbance was present.

Model Modification

The SIMPLIS program was rerun with the following added commands to correlate the error covariances:

```
Let error covariance between age11 and age12 correlate
Let error covariance between age14 and age15 correlate
```

After modification, the latent growth curve model had a more acceptable model fit (chi-square = 11.35, $df = 5$, and $p = .05$). The final latent growth curve model output with standardized coefficients is diagrammed in Figure 14.2. The individual slopes *increased* over time as indicated in Table 14.1.

The intercepts *decreased* over time as indicated in Table 14.2.

The negative correlation between the slope and intercept correctly indicated the increase in slope values over time with a corresponding decrease in intercept values over time ($r = -.53$).

Table 14.1: Slope Values by Age

Group	Slope
Age 11	.00
Age 12	.23
Age 13	.35
Age 14	.49
Age 15	.62

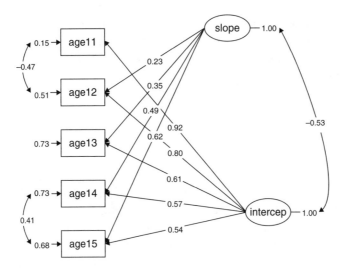

Chi–Square=11.49, df = 5, P-value=0.04250, RMSEA=0.088

■ **Figure 14.2:** MODIFIED LATENT GROWTH CURVE MODEL (STANDARDIZED SOLUTION)

Note: The SIMPLIS computer output does not list the slope and intercept values, but does display them in the path model diagram. They were copied and listed above for convenience.

A test of the linear rate of growth in the latent growth curve model seemed appropriate because the means increased from .20 at age 11 to .45 at age 15. The latent growth curve model is appropriately called a *Latent Growth Curve Structured Means Model* because group means as well as covariance were specified. There were individual differences in the slopes over time. The negative correlation between the intercept values (group means) and the slope values (linear growth) indicated that as age increased the level of tolerance decreased.

Table 14.2: Intercept Values by Age

Group	Intercept
Age 11	.92
Age 12	.80
Age 13	.61
Age 14	.57
Age 15	.54

This LGM model indicated a linear rate of growth in adolescent tolerance for deviant behavior using age 11 as the baseline for assessing linear change over time. You should graph these mean values across the age levels to graphically display the trend. You should also interpret the correlation between the intercept and slope because a positive value would indicate different results, namely that high initial status at age 11 had a greater rate of change, while a negative correlation would indicate that high initial status at age 11 had a lower rate of change. If the average slope value is zero, then no linear change has occurred. Finally, you can assess how measurement errors across adjacent years are correlated (lagged correlation in ARIMA models). This ability to model measurement error is a unique advantage of LGM over traditional ANOVA repeated measure designs.

LATENT GROWTH CURVE MODEL—GROUP DIFFERENCES

In the previous LGM model, individual change over time was indicated by the intercept and slope, which was similar to a one-factor (one-way) repeated measures analysis of variance. We can extend the LGM model to include group differences. These group differences can be fixed, for example, race or gender, or they could be time-varying, for example, memory retention, cognitive ability, or weight. The covariates can also be fixed or continuous variables. Our example tests group differences (gender) as a dummy coded variable (1 = male; 0 = female) across four fixed time points (3, 6, 9, 12 months) on a numerical rating of cell phone customer satisfaction for 35 customers. The LGM is diagrammed in Figure 14.3, and indicates parameter estimates for slope (a), intercept (b), and the dummy coded group difference variable (D1). The error terms for slope (c) and intercept (d) are correlated using the curved arrow. The error covariances (e3, e6, e9, e12) are uncorrelated. The slope coefficients (0, 1, 2, 3) indicate a test of linear trend across the 3- to 12-month time periods. The intercept coefficients (1, 1, 1, 1) indicate the means for the time periods.

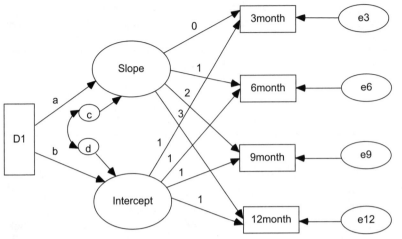

■ Figure 14.3:   LATENT GROWTH CURVE MODEL—GENDER DIFFERENCES (AMOS)

The level 1 model can be expressed as

$$Y_{iT} = \alpha_i + \lambda_T \beta_i + \varepsilon_{iT}.$$

The equation indicates that numerical ratings (Y) are a function of the intercept (α) and slope (β) for the time periods (T). The level 2 model, which computes the intercepts (α) and slopes (β), can be expressed as

$$\alpha_i = \mu_\alpha + \gamma_{\alpha D1} D_{1i} + \varsigma_{\alpha i}$$

and

$$\beta_i = \mu_\beta + \gamma_{\beta D1} D_{1i} + \varsigma_{\beta i}.$$

In Figure 14.3, a = $\gamma_{\beta D1} D_{1i}$, which is the slope parameter estimate for the dummy coded variable (Male = 1); b = $\gamma_{\alpha D1} D_{1i}$, which is the intercept parameter estimate for the dummy coded variable (Male = 1); with c = $\varsigma_{\beta i}$ and d = $\varsigma_{\alpha i}$, which are correlated residual error terms for the slope and intercept equations, respectively.

The *lavaan* package in R was used to run the LGM model to test for group difference in customer satisfaction across the four time periods. The R program and data set is on the book website (LGM.*r*; satisfied.txt). The R commands in the program are explained as follows:

1. Read in data set and print first 6 lines of data

```
LGMdata = read.table(file="c:/satisfied.txt",header=TRUE,sep=",")
```

```
head(LGMdata)
 id y1 y2 y3 y4 gender
1 1 2.5 2.5 3.0 4.5    0
2 2 1.0 2.3 2.9 5.0    0
3 3 1.5 1.7 1.6 2.9    1
4 4 0.5 2.0 2.7 3.9    1
5 5 0.0 1.7 3.1 3.6    1
6 6 0.5 1.9 3.7 6.9    0
```

2. Model specification and naming of LGM model in *Lavaan*

```
satisfy = '
# intercept
i = ~ 1*y1 + 1*y2 + 1*y3 + 1*y4
# slope
s = ~ 0*y1 + 1*y2 + 2*y3 + 3*y4
# regression
i + s = ~ gender
'
```

The model is named *satisfy*, and placed in single quotation marks. The intercept (i) means are computed in the equation using 1;s before each observed variable (y1 to y4). The slope(s) are computed in the equation using the linear coefficients (0, 1, 2, 3) before each observed variable. Finally, the intercept (i) and slope (s) are predicted in a regression equation with *gender*, the dummy coded variable.

3. The LGM model is run using the growth() function in *Lavaan*; results output.

```
modelfit = growth(satisfy,data=LGMdata)
summary(modelfit)
```

The results indicated a good data to model fit ($\chi^2 = 14.24$, $df = 8$, $p = .08$). The intercepts were not different for males and females ($z = .875$, $p = .382$), with the male value reported as 4.341; standard error = 4.962. The slopes were not different for the males and females ($z = -0.817$, $p = 0.414$), with the male value reported as -5.101; standard error = 6.246. Finally, intercepts ($z = 17.957$, $p < 0.0001$) and slopes ($z = 15.094$, $p < 0.0001$) differed across the four time periods, but not due to gender differences. The computer output is shown below.

Computer Output

	Estimate	Std.err	Z-value	P(>|z|)
Number of observations				35
Estimator				ML
Minimum Function Test Statistic				14.240
Degrees of freedom				8
P-value (Chi-square)				0.076
Parameter estimates:				
Information				Expected
Standard Errors				Standard
	Estimate	Std.err	Z-value	P(>|z|)
i =~				
gender	4.341	4.962	0.875	0.382
s =~				
gender	-5.101	6.246	-0.817	0.414
Covariances:				
i ~~				
s	0.078	0.022	3.538	0.000
Intercepts:				
i	1.379	0.077	17.957	0.000
s	1.079	0.071	15.094	0.000
Variances:				
y1	0.283	0.069		
y2	0.046	0.027		
y3	0.309	0.094		
y4	0.873	0.241		
gender	0.017	0.239		
i	0.071	0.040		
s	0.090	0.039		

We can substitute these values into the level 2 equations as follows:

$$\alpha_i = \mu_\alpha + \gamma_{\alpha D1} D_{1i} + \varsigma_{\alpha i}$$
$$\alpha_i = 1.379 + 4.34 D_{1i} + \varsigma_{\alpha i}$$

and

$$\beta_i = \mu_\beta + \gamma_{\beta D1} D_{1i} + \varsigma_{\beta i}$$
$$\beta = 1.079 - 5.10 D_{1i} + \varsigma_{\beta i}$$

This shows that the male intercept was slightly higher than the female intercept on their mean satisfaction at time 1. The male slope value (4.34) indicates that males increased more than females (−5.10) over the four time periods. The analysis, however, indicated **no** gender difference in the intercept and slopes. A plot of the intercept values over the four time periods in Figure 14.4 provides a visual display for each group. A common slope (middle line) is also plotted to provide

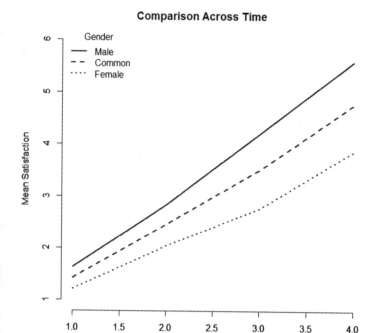

■ **Figure 14.4:** PLOT OF MEAN SATISFACTION BY GENDER

a visual display of the increase in intercepts and common slope, without a group difference.

SUMMARY

This chapter presented two different latent growth curve models. The first model only included a single factor across time periods. This is similar to one-way repeated measures in analysis of variance. We examined the mean (intercept) and rate of change (slope) across age groups on the response variable, tolerance for deviant behavior. Results indicated that as age increased, tolerance for deviant behavior decreased. The second latent growth curve model extended the intercept and slope model to include a group membership variable. This would be considered a mixed design in repeated measures analysis of variance where the researcher is determining within and between effects. The within effects are the intercept and slope changes over time, while the between effect is the group difference across the time periods. Our results indicated no gender difference in the intercept and slope values across time. We were able to show a change across time, but not due to gender differences.

There are many SEM software programs that conduct latent growth curve modeling. Examples are presented in numerous books. Bollen and Curran (2006) show several examples using AMOS, Mplus, and SPSS. Duncan, Duncan, Strycker, Li, and Alpert (1999) provide numerous examples using SPSS and EQS syntax programs to illustrate various types of models. Acock (2013) analyzes latent growth curve models using Stata. Byrne (2012) includes a chapter using Mplus to analyze latent growth curve models. Raykov and Marcoulides (2000) include a chapter using LISREL and EQS to analyze latent change. Heck, Thomas, and Tabata (2010) use numerous SPSS mixed model procedures to demonstrate how to conduct longitudinal modeling with IBM SPSS software. Muthén and Muthén (2006) provide Mplus examples in many different types of models using categorical and continuous covariates, fixed or random intercepts and slopes, and dummy or continuous variables.

EXERCISES

1. LGM—Intercept and Slope

News and radio stations in Dallas, Texas, have convinced the public that a massive crime wave has occurred during the past 4 years, from 2002 to 2005. A criminologist gathered the crime rate data, but needs your help to run a latent growth curve model to test whether a linear trend in crime rates exists for the city. The data set information is:

```
Observed variables: time1 time2 time3 time4
Sample Size 400
Correlation Matrix
1.000
 .799 1.000
 .690 .715 1.000
 .605 .713 .800 1.000
Means 5.417 5.519 5.715 5.83
Standard Deviations .782 .755 .700 .780
```

Create an SEM program, diagram the model with standardized coefficients, and interpret your findings. Have crime rates increased in Dallas from 2002 to 2005?

2. LGM—Group Difference

The data set *region.txt* contains the number of tornado sightings in Oklahoma and Virginia for 3 months, sighted by 10 individuals in each state. Region is dummy coded (1 = Oklahoma; 0 = Virginia). Run the R program to determine if the intercepts and/or slopes differ for each region. Alternatively, run the LGM in a different SEM software program. Report the values.

```
# Read in Data set
LGdata = read.table(file="c:/region.txt",header=TRUE,sep=",")
head(LGdata)
# lavaan SEM package
install.packages("lavaan")
library(lavaan)
# LG model
sighting = '
# intercept
i = ~ 1*y1 + 1*y2 + 1*y3
# slope
s = ~ 0*y1 + 1*y2 + 2*y3
# regression
i + s = ~ region
'
# Run LGM
model = growth(sighting,data=LGdata)
summary(model)
```

SUGGESTED READINGS

Duncan, S. C., & Duncan, T. E. (1994). Modeling incomplete longitudinal substance use data using latent variable growth curve methodology. *Multivariate Behavioral Research, 29*(4), 313–338.

Ghisletta, P., & McArdle, J. J. (2001). Latent growth curve analyses of the development of height. *Structural Equation Modeling: A Multidisciplinary Journal, 8*, 531–555.

Shevlin, M., & Millar, R. (2006). Career education: An application of latent growth curve modeling to career information-seeking behavior of school pupils. *British Journal of Educational Psychology, 76*, 141–153.

REFERENCES

Acock, A. C. (2013). *Discovering structural equation modeling using Stata* (revised edition). College Station, TX: Stata Press.

Bollen, K. A., & Curran, P. J. (2006). *Latent curve models: A structural equation modeling perspective.* Wiley Series in Probability and Statistics. New York: John Wiley & Sons.

Byrne, B. M. (2012). *Structural equation modeling with Mplus.* Multivariate Applications Series. New York: Routledge.

Byrne, B. M., & Crombie, G. (2003). Modeling and testing change: An introduction to the latent growth curve model. *Understanding Statistics*, 2(3), 177–203.

Duncan, T. E., & Duncan, S. C. (1995). Modeling the processes of development via latent variable growth curve methodology. *Structural Equation Modeling*, 2(3), 187–213.

Duncan, T. E., Duncan, S. C., Strycker, L. A., Li, F., & Alpert, A. (1999). *An introduction to latent variable growth curve modeling: Concepts, issues, and applications.* Mahwah, NJ: Lawrence Erlbaum Associates.

Duncan, T. E., Duncan, S. C., & Strycker, L. A. (2006). *An introduction to latent variable growth curve modeling: Concepts, issues, and applications* (2nd edn.). Mahwah, NJ: Lawrence Erlbaum Associates.

Heck, R. H., Thomas, S. L., & Tabata, L. N. (2010). *Multilevel and longitudinal modeling with IBM SPSS.* Quantitative Methodology Series. New York: Taylor & Francis.

Li, F., Duncan, T. E., Duncan, S. C., Acock, A. C., Yang-Wallentin, F., & Hops, H. (2001). Interaction models in latent growth curves. In G. A. Marcoulides, & R. E. Schumacker (Eds.), *New developments and techniques in structural equation modeling* (pp. 173–201). Mahwah, NJ: Lawrence Erlbaum.

Lomax, R. G., & Hahs-Vaughn, D. L. (2012). *An introduction to statistical concepts* (3rd edn.). New York: Routledge.

McArdle, J. J., & Epstein, D. (1987). Latent growth curves within developmental structural equation models. *Child Development*, 58, 110–133.

Muthén, L. K., & Muthén, B. O. (2006). *Mplus user's guide* (4th edn.). Los Angeles: Muthén & Muthén.

Raykov, T., & Marcoulides, G. A. (2000). *A first course in structural equation modeling.* Mahway, NJ: Lawrence Erlbaum Associates.

Schumacker, R. E. (2016). *Using R with multivariate statistics.* Thousand Oaks, CA: Sage.

Stoolmiller, M. (1995). Using latent growth curves to study developmental processes. In J. M. Gottman (Ed.), *The analysis of change* (pp. 103–138). Mahwah, NJ: Lawrence Erlbaum.

Chapter 15

SEM INTERACTION MODELS

CHAPTER CONCEPTS

Main effects and interaction effects
Kenny–Judd approach
Latent variable interaction model
Latent variable Mplus program
Computing latent variable scores
Computing latent interaction variable
Model modification
Two-Stage Least Squares (TSLS) approach

MAIN EFFECTS AND INTERACTION EFFECTS

Researchers have focused more on testing main effects than interaction effects in regression and path models (Newman, Marchant, & Ridenour, 1993). In fact, structural equation modeling began as a way of testing linear structural relations (LISREL). It wasn't until Kenny and Judd proposed interaction amongst observed variables in a model, that interaction effects began to be noticed (Kenny & Judd, 1984; Bollen, 1989). Today, the categorical, continuous, and non-linear interaction approaches to structural equation modeling have become noticed and further explored (Schumacker & Marcoulides, 1998; Yang-Wallentin & Jöreskog, 2001).

Most SEM models today still assume that the relations in the models are linear, that is, the relations among all variables, observed and latent, are represented by linear equations. Several studies have been published where non-linear and interaction effects are used in multiple regression models; however, these effects have seldom been tested in path models, and you will infrequently find non-linear factor models. It should not be surprising to find that for several decades structural equation modeling was based on solving a set of simultaneous linear structural equations.

SEM models with non-linear and interaction effects are now possible and can easily be modeled with recent versions of SEM software. However, there are several types of non-linear and interaction effects: categorical, product indicant, non-linear, two-stage least squares, and latent variable using normal scores. For continuous observed variables, a non-linear relationship could exist between two observed variables (X_1 and X_2 are curvilinear); a quadratic (non-linear) term in the model ($X_2 = X^2_1$); or a product of two observed variables ($X_3 = X_1X_2$). These three different types of interaction effects all involve *continuous* observed variables. For *categorical* observed variables, interaction effects are similar to analysis of variance and use the multiple-group SEM model (Schumacker & Rigdon, 1995). These continuous variable and categorical variable approaches also apply to latent variables.

Given that so many different approaches exist, this chapter will cover the Kenny–Judd approach of using the interaction of observed variables in a model, and the newer approach of using latent variables to compute interactions in a model. The latent variable interaction uses the product of individual latent variable scores that are computed and added to the data file from the measurement models. We encourage you to also run models available in your SEM user's guide.

KENNY–JUDD APPROACH

Kenny and Judd (1984) presented the method of non-linear and interactive effects using latent variables. They coined the term *product indicators*, when introducing an observed variable that was squared into the equation, or when multiplying two observed variables. These variables would then be used as indicators of a latent variable. Bollen (1989) further explained the Kenny–Judd approach. The basic product indicant approach is diagrammed in Figure 15.1. The independent latent variable (Ksi) has VAR1, VAR2, and VAR12 (interaction) as indicators. The dependent latent variable (Eta) has VAR3, VAR4, and VAR5 as indicator variables. You could also include a product indicator interaction term for the Eta latent variable. *Note:* The error variances for the indicator variables are not shown.

SIMPLIS Program

The data for the SIMPLIS program are in KJUDD.LSF. It originally contained five variables; the first two are VAR1 and VAR2, and the remaining three variables are VAR3, VAR4, VAR5. The product indicator variable, VAR12, was added to

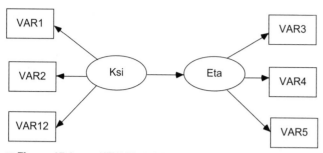

■ Figure 15.1:   KENNY–JUDD INTERACTION MODEL (AMOS)

the file and created prior to running the program by using the compute statement (Transformation on pull-down menu of LISREL9). Output from the PRELIS program is automatically generated when computing a new variable. The output indicates a sample size of 1000, the univariate summary statistics, and the univariate normality tests for each variable. *Note*: The VAR1, VAR2, VAR3, VAR4, and VAR5 variables should have a mean = 0 and standard deviation = 1 (from the randomly generated data). Also, given simulated random normal data, the skewness and kurtosis values should be non-significant. The results in Table 15.1, however, show that VAR1 did not have a mean = 1, and the skewness and kurtosis values were statistically significant (χ^2 = 704.912), which was carried over to the product indicator variable (VAR12; χ^2 = 14,306.32).

Table 15.1: Descriptive Statistics (N = 1000)

Univariate Summary Statistics for Continuous Variables

Variable	Mean	St. Dev.	Skewness	Kurtosis	Minimum	Freq.	Maximum	Freq.
VAR1	1.181	0.796	1.208	3.612	−1.290	1	5.680	1
VAR2	0.042	1.001	0.010	−0.097	−3.170	1	3.440	1
VAR3	−0.003	0.914	−0.137	0.074	−3.960	1	2.440	2
VAR4	0.018	0.975	0.014	0.019	−3.420	1	3.080	1
VAR5	−0.006	0.904	0.054	−0.011	−2.760	1	3.080	1
VAR12	0.262	1.703	2.263	18.259	−6.780	1	19.023	1

Test of Univariate Normality for Continuous Variables

	Skewness		Kurtosis		Skewness and Kurtosis	
Variable	Z-Score	P-Value	Z-Score	P-Value	Chi-Square	P-Value
VAR1	12.597	0.000	23.371	0.000	704.912	0.000
VAR2	0.127	0.899	−0.626	0.531	0.408	0.815
VAR3	−1.772	0.076	0.480	0.631	3.370	0.185
VAR4	0.181	0.856	0.124	0.901	0.048	0.976
VAR5	0.698	0.485	−0.074	0.941	0.493	0.781
VAR12	18.596	0.000	118.155	0.000	14306.320	0.000

The SIMPLIS program to analyze the data given the model in Figure 15.1 was written as:

```
Kenny Judd Model
Raw Data from File KJUDD.LSF
Latent variables Ksi Eta
Relationships
VAR1 VAR2 VAR12 = Ksi
VAR3 VAR4 VAR5 = Eta
Eta = Ksi
Path Diagram
End of Problem
```

The results indicated a poor model fit ($\chi^2 = 258.40$, $df = 8$, $p > .000001$). VAR3 was not a statistically significant indicator of Eta (dependent latent variable), and error covariances between VAR1 and VAR2, and VAR2 and VAR12 needed to be added. We therefore dropped VAR3 from the model, added the error covariance terms, and re-ran the program. The new SIMPLIS program was written as:

```
Kenny Judd Model
Raw Data from File KJUDD.LSF
Latent variables Ksi Eta
Relationships
VAR1 VAR2 VAR12 = Ksi
VAR4 VAR5 = Eta
Eta = Ksi
! Drop VAR3 - non-significant, add error covariance terms
Let the error covariance of VAR1 and VAR2 correlate
Let the error covariance of VAR2 and VAR12 correlate
Path Diagram
End of Problem
```

This final model indicated a good data to model fit ($\chi^2 = 5.63$, $df = 2$, $p = .06$, RMSEA = .04). The model in Figure 15.2 shows the standardized parameter estimates.

The parameter estimates in the model were all statistically significant. The following computer output provided the parameter estimates, standard errors, z values, and p values for the measurement models (Ksi and Eta), and the structural model where Ksi predicted Eta (dependent latent variable).

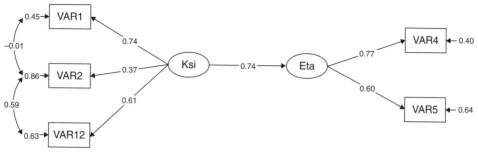

■ Figure 15.2: KENNY–JUDD MODEL (MODIFIED—LISREL PATH DIAGRAM)

SIMPLIS Output

```
LISREL Estimates (Maximum Likelihood)
          Measurement Equations
      VAR4 = 0.755*Eta, Errorvar.= 0.380, R² = 0.600
Standerr                      (0.0458)
Z-values                       8.297
P-values                       0.000


      VAR5 = 0.539*Eta, Errorvar.= 0.527, R² = 0.355
Standerr   (0.0446)           (0.0320)
Z-values   12.079             16.457
P-values    0.000              0.000


      VAR1 = 0.589*Ksi, Errorvar.= 0.286, R² = 0.549
Standerr   (0.0309)           (0.0292)
Z-values   19.087             9.809
P-values    0.000              0.000


      VAR2 = 0.374*Ksi, Errorvar.= 0.861, R² = 0.140
Standerr   (0.0445)           (0.0456)
Z-values    8.405             18.888
P-values    0.000              0.000


     VAR12 = 1.035*Ksi, Errorvar.= 1.830, R² = 0.369
Standerr   (0.0624)           (0.115)
Z-values   16.597             15.916
P-values    0.000              0.000


Error Covariance for VAR2 and VAR1 = -0.008
                                    (0.0181)
                                    -0.445
```

```
        Error Covariance for VAR12 and VAR2 = 1.003
                                            (0.0678)
                                             14.793
                    Structural Equations


        Eta = 0.744*Ksi, Errorvar.= 0.447, R² = 0.553
  Standerr    (0.0477)                    (0.0808)
  Z-values    15.604                       5.532
  P-values     0.000                       0.000
```

Summary

Jöreskog and Yang (1996) recommend adding means to variables in the Kenny–Judd model. The means of the indicator variables are known to be a function of other parameters in the model, so their intercepts should be added to yield more accurate results. They also suggested that a single product indicator is sufficient for model identification. Marsh, Wen, and Hau (2004) suggested a compromise between single product indicators and all-possible product indicators by suggesting a matched-pair product indicator approach in a bi-factor measurement model. This essentially involves X1 and X2 as indicators of F1, X3 and X4 as indicators of F2, with them being cross-multiplied (X1 * X3; X2*X4) to create the matched pair product indicators.

Our example did not input mean values for the indicator variables because they were standardized with mean = 0 and standard deviation = 1 (based on randomly generated data). Another issue that surrounds the use of product indicant variables has centered on the normal distribution assumption. Basically, X1 and X2 indicator variables can be normally distributed, but X1*X2 interaction can be non-normally distributed. We saw this happen in our Kenny–Judd data example. Also, a product indicator variable (X3 = X1*X2) will generally have severe multicollinearity with the individual X1 and X2 observed variables. Another approach using the product of latent variables has been suggested (Bollen, 1989; Wong & Long, 1987), and will be covered next.

LATENT VARIABLE INTERACTION MODEL

Bollen (1989) presented the matrix algebra behind a model that included the product of two latent variables to indicate a latent variable interaction term. A latent variable interaction model would hypothesize that the independent latent variables (*Ksi1* and *Ksi2*) would be multiplied to obtain the product of *Ksi1* * *Ksi2*, to yield *Ksi12*. The model would include these independent latent variables in the prediction of a dependent latent variable (*Eta*). The latent variable interaction model is diagrammed in Figure 15.3. *Note:* The observed indicator variables of the latent variables have been omitted in the latent variable interaction model diagram.

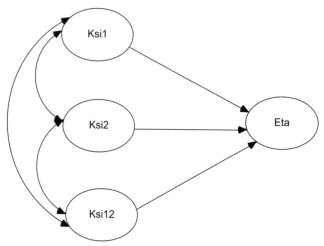

■ **Figure 15.3:**   LATENT VARIABLE INTERACTION MODEL

Latent Variable Program

A data set, *kjint.txt*, was generated and used in the following latent variable interaction Mplus program. The x1 and x2 observed variables are indicators of the latent variable, X. The z1 and z2 observed variables are indicators of the latent variable, Z. These two latent variables are then multiplied to compute, $X \times Z$ (latent variable interaction). Finally, the dependent variable y is regressed on X, Z, and $X \times Z$ latent variables. The Mplus program (with a few changes) was written to input the raw data, conduct the measurement models, compute the latent variable interaction, and then analyze the relations amongst the latent variables in the interaction model. The Mplus program is followed by the Mplus results (abbreviated output), and the Mplus model diagram.

Mplus Program (Modified)

```
TITLE: latent Variable interaction example Data generated by Phil Wood.
DATA: FILE IS kjint.txt;
VARIABLE: NAMES ARE y x1 x2 z1 z2;
ANALYSIS: TYPE = RANDOM;!required to run interaction effects
ALGORITHM = INTEGRATION;
MODEL: X BY x1 x2;
Z by z1 z2;
```

```
XxZ | X XWITH Z;
y ON X Z XxZ;
OUTPUT: SAMPSTAT;! no standardized estimates with type = random
plot: type=plot3;
```

MPlus Output (Abbreviated)

The SAMPSTAT option permits knowledge that x1 and x2 are not correlated with z1 and z2 in the model. This is not usually the case, but apparent here from the simulated data set by Phil Wood.

```
        Correlations
           Y       X1      X2      Z1      Z2


   Y     1.000
   X1    0.083   1.000
   X2    0.033   0.647   1.000
   Z1    0.187   0.000   0.000   1.000
   Z2    0.193   0.000   0.000   0.649   1.000
THE MODEL ESTIMATION TERMINATED NORMALLY
MODEL FIT INFORMATION
Number of Free Parameters         18
Loglikelihood
        H0 Value                       -3727.410
        H0 Scaling Correction Factor    1.0340
        for MLR
Information Criteria
        Akaike (AIC)                    7490.819
        Bayesian (BIC)                  7566.682
        Sample-Size Adjusted BIC        7509.549
          (n* = (n + 2) / 24)
MODEL RESULTS
                                             Two-Tailed
             Estimate    S. E.   Est./S. E.   P-Value
   X BY
      X1        1.000     0.000    999.000     999.000
      X2        0.620     0.032     19.609       0.000
   Z BY
      Z1        1.000     0.000    999.000     999.000
      Z2        0.734     0.043     17.190       0.000
   Y ON
      X         0.132     0.044      2.990       0.003
      Z         0.337     0.048      7.033       0.000
      XXZ       0.711     0.032     22.538       0.000
   Z WITH
      X         0.097     0.096      1.012       0.312
```

```
Intercepts
    Y           -0.069    0.037    -1.872    0.061
    X1           0.000    0.069     0.004    0.997
    X2           0.000    0.055     0.003    0.998
    Z1           0.000    0.063    -0.003    0.998
    Z2           0.000    0.052    -0.002    0.998
Variances
    X            2.031    0.161    12.639    0.000
    Z            1.537    0.133    11.538    0.000
Residual Variances
    Y            0.117    0.025     4.688    0.000
    X1           0.477    0.068     7.027    0.000
    X2           0.804    0.058    13.834    0.000
    Z1           0.551    0.070     7.897    0.000
    Z2           0.596    0.050    11.887    0.000
```

Mplus Diagram

Figure 15.4 shows the latent variable interaction model with the indicator variables for the X and Z latent variables. It also shows the latent variable $X \times Z$ created by multiplying these two latent variables. Finally, the two main effect latent variables (X and Z), and the interaction effect latent variable ($X \times Z$) are shown predicting the dependent variable, Y. *Note*: The dependent variable is not a latent variable. The Mplus diagram also prints the unstandardized parameter estimates and the standard errors in parentheses. Recall, the t or z value is computed as the parameter estimate divided by the standard error, which is shown in the computer output above.

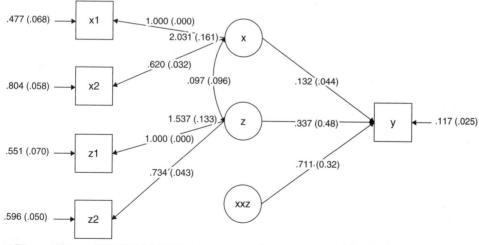

■ **Figure 15.4:** LATENT VARIABLE INTERACTION MODEL (MPLUS DIAGRAM)

Computing Latent Variable Scores

Latent variable scores were first introduced by Jöreskog (2000) and Jöreskog et al. (2000). The issue of latent variable scores and their use in creating an interaction term was investigated by Klein and Moosbrugger (2000). Basically, the two variables should be *centered*, which is computed by subtracting the variable value from its mean, which computes a deviation score. The centered variables are then used to compute the product variable. A discussion on the importance of centering is given by Aiken and West (1991).

Computing Latent Interaction Variable

You create the latent interaction variable by multiplying the latent variable scores *ksi1* and *ksi2*. These latent variable scores are unbiased and produce the same mean and covariance matrix as the latent variables.

Model Modification

If the parameter estimate for the interaction latent variable is not statistically significant, simply drop the interaction latent variable and run the model using only *ksi1* and *ksi2*.

Two-stage Least Squares (TSLS) Approach

Recent developments in non-linear structural equation modeling have focused on full information methods (maximum likelihood, or asymptotically distribution free methods) because of a concern about estimating unbiased parameters and standard errors. We recommend using bootstrap estimates of the parameter estimates and standard errors in non-linear models, because the observed and/ or latent interaction variables generally do not meet the multivariate normality assumption. Other problems or sources of error could exist that hinder convergence to a solution, which is why *start values* are recommended to aid convergence, for example, use initial 2SLS estimates or user-defined start values. The two-stage least-squares (2SLS) estimates and their standard errors are obtained without iterations and therefore provide the researcher with clues to which parameters exceed their expected values (correlations with values greater than 1.0 in a non-positive definite matrix). The 2SLS estimates therefore provide helpful information to determine whether the specified model is reasonable (Bollen, 1995, 1996).

SUMMARY

SEM interaction models comprise many different types of models. The use of continuous variables, categorical variables, non-linear effects, and latent variables in interaction models have intrigued scholars over the years. The current approaches that appear easy to model are the multi-group categorical approach and the latent variables score approach, because they are not affected by many of the problems related to multivariate normality and centering.

The testing of interaction effects can present problems in structural equation modeling. First, you may have the problem of model specification. Linear models simplify the task of determining relations to investigate and distributional assumptions to consider, but this may not be the case in latent variable interaction models. Second, discarding the linearity assumption opens up the possibility of several product indicant variable and latent variable interaction combinations, but this also serves to magnify the critical role of theory in focusing the research effort. Third, a researcher who seeks to model categorical interaction effects must also collect data that span the range of values in which interaction effects are likely to be evident in the raw data, and must collect a sample size large enough to permit subsamples. Fourth, we have noted that the statistical fit index and parameter standard errors are based on linearity and normality assumptions, and we may not have robust results to recognize the presence of an interaction effect unless it is substantial.

The continuous variable approach does have its good points. It is possible to check for normality of variables, and to standardize them (normal scores), and the approach does not require creating subsamples or forming groups where observations could be misclassified, nor does it require the researcher to categorize a variable and thereby lose information. Moreover, the continuous variable approach is parsimonious. Basically, all but one of the additional parameters involved in the interaction model are exact functions of the main-effects parameters, so the only new parameters to be estimated are the structure coefficient for the latent interaction independent variable and the prediction equation error.

The continuous variable approach also has several drawbacks. First, only a few software programs can perform the necessary non-linear constraints, and the programming for testing interaction effects in the traditional sense is not easy. Second, if you include too many indicator variables of your latent independent variables, this approach can become very cumbersome. For example, if one latent independent variable, Factor 1, has n_1 measures and the other latent independent variable, Factor 2, has n_2 measures, then the interaction term, Factor 1 × Factor 2, could have $n_1 \times n_2$ measures. If each independent latent variable has five indicator variables, then the multiplicative latent independent variable interaction would involve 25 indicators. Including the 5 measures for each of the 2 main-effect latent independent variables and 2 indicators of a latent dependent variable, the model would have 37 indicator variables before any other latent-variable relationships were considered. Third, the functional form of the interaction needs to be specified. The simple multiplicative interaction presented in this chapter hardly covers other types of interactions, and for these other types of interactions there is little prior research or available examples to guide the researcher.

A fourth problem to consider is multicollinearity. It is very likely that the interaction factor will be highly correlated with the observed variables used to construct it. This multicollinearity in the measurement model causes the interaction latent independent variable to be more highly correlated with the observed variables of other main effect latent independent variables than each set of observed variables are with their own respective main effect independent variables. For multiplicative interactions between normally distributed variables, multicollinearity could be eliminated by centering the observed variables (using scores expressed as deviations from their means) before computing the product variable. However, centering the variables alters the form of the interaction relationship. Researchers who want to model other types of interactions may find no easy answer to the problem of multicollinearity (Smith & Sasaki, 1979).

A fifth concern relates to distributional problems, which are more serious than those associated with linear modeling techniques using only observed variables. If the observed variables are non-normal, then the variance of the product variable can be very different from the values implied by the basic measurement model, and the interaction effect will perform poorly. Of course, permissible transformations may result in a suitable, normal distribution for the observed variables. The resultant non-normality, however, in the observed variables violates the distributional assumptions associated with the estimation methods used, for example, maximum likelihood. Furthermore, estimation methods that do not make distributional assumptions may not work for interaction models.

When using the latent variable score approach you should consider bootstrapping the standard errors because the estimation method used may give inaccurate estimates of standard errors given violation of the distributional assumption for the interaction model. Basically, the asymptotic weight matrix associated with the covariance matrix for an interaction model may be non-positive definite because of dependencies between moments of different observed variables that are implied by the interaction model. In any case, we would recommend that you bootstrap the parameter estimates and standard errors to achieve a more reasonable estimate of these values (Bollen & Stine, 1993; Jöreskog & Sörbom, 1993; Lunneborg, 1987; Mooney & Duval, 1993; Stine, 1990; Yang-Wallentin & Jöreskog, 2001).

The use of non-linear and interaction effects is popular in regression models (Aiken & West, 1991; Jaccard & Turrisi, 2003). The inclusion of interaction hypotheses in path models has been minimal (Newman, Marchant, & Ridenour, 1993), and few examples of non-linear factor models have been provided (Etezadi-Amoli & McDonald, 1983; McDonald, 1967). SEM models with interaction effects are now possible and better understood due to several scholars including Kenny and Judd (1984), Wong and Long (1987), Bollen (1989), Higgins and Judd (1990), Cole, Maxwell, Arvey, and Salas (1993), Mackenzie and Spreng (1992), Ping (1993, 1994, 1995), Jöreskog and Yang (1996), Schumacker and Marcoulides (1998), Algina and Moulder (2001), du Toit and du Toit (2001), Moulder and Algina (2002), and Schumacker (2002), to name only a few.

Jöreskog and Yang (1996) provided additional insights into modeling interaction effects, given the problems and concerns discussed here. Jöreskog (2000) discussed many issues related to interaction modeling and included latent variable scores in LISREL that are easy to compute and include in interaction modeling. Schumacker (2002) compared the latent variable score approach to the continuous variable approach using the LISREL matrix command language and found the parameter estimates to be similar with standard errors reasonably close. Our recommendation would be to use the latent variable score approach and bootstrap the standard errors. If unfamiliar with the bootstrap approach, then use normal scores with interaction variables to avoid non-normal issues when testing interaction effects.

Structural equation models that include interaction effects are not prevalent in the research literature, in part because of all the concerns mentioned here. The categorical variable approach using multiple samples and constraints has been used most often. The latent variable score approach using normal scores is a useful way to model interaction with latent variables. Hopefully, more SEM research will consider interaction hypotheses given the use of latent variable scores and the use of *normal score* data for main effect and interaction variables.

EXERCISES

1. Kenny–Judd Product Indicant Model

Figure 15.5 shows three observed indicator variables (V11, V12, V13) and an interaction product indicator variable (V17_21) for *Self*. The CFA model is testing whether the interaction of Age and Stress affects Sense of Self (Self). The variables used in the data were: V11 = Self-Esteem; V12 = Locus of Control: V13 = Marital

Satisfaction; V17 = Age; V21 = Life Change (Stress). The interaction variable was created as: V17_21 = V17 * V21.

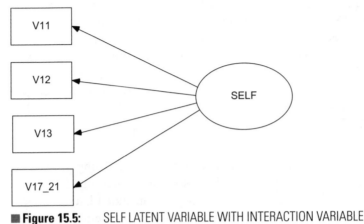

■ **Figure 15.5:** SELF LATENT VARIABLE WITH INTERACTION VARIABLE

The example is a subset of an analysis by Tabachnick and Fidell (2007, 5th edition).

The data were downloaded from the Internet, and missing values (0) were inserted, $N = 459$. Data: media.pearsoncmg.com/ab/ab_tabachnick_multistats_6/datafiles/ASCII/hlthsem.dat. The subset of data used in the exercise is available on the book website in the file, *health.txt*.

Write an SEM program to test the statistical significance of the parameter estimates (paths) in the model. Did the data fit the model? Were all parameter estimates statistically significant? Specifically, was the interaction variable, V17_21, statistically significant?

2. Latent Variable Interaction Model

An organizational psychologist was investigating whether the independent latent variables, *work tension* and *collegiality*, were predictors of a dependent latent variable, *job satisfaction*. Research indicated that *work tension* and *collegiality* interact, so an interaction model was hypothesized and tested. The model is diagrammed in Figure 15.6. The data file contains $n = 200$ employees' responses on 9 variables (*jobs.txt*). Open data in LISREL, then write a program to create the latent variables, and the interaction variable. Check that they have been added to the LISREL system file. Next, create and run a SIMPLIS program to test the interaction model. What conclusions can you make regarding the interaction of the latent variables *work tension* and *collegiality*?

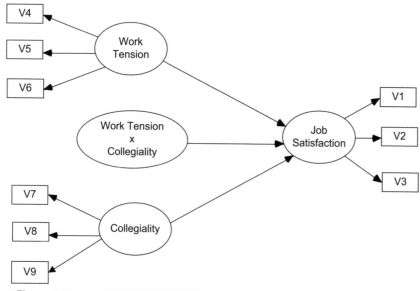

■ **Figure 15.6:** JOB SATISFACTION INTERACTION MODEL (AMOS)

SUGGESTED READINGS

Fielding, D., & Torres, S. (2005). A simultaneous equation model of economic development and income inequality. *Journal of Economic Inequality*, 4, 279–301.

Hancock, G. R., & Liu, M. (2012). Bootstrapping standard errors and data-model fit statistics. In R. Hoyle (Ed.), *Handbook of structural equation modeling* (pp. 296–306). New York: Guilford Press.

Harring, J. R., Weiss, B. A., & Hsu, J.-C. (2012). A comparison of methods for estimating quadratic effects in nonlinear structural equation models. *Psychological Methods*, 17(2), 193–214.

Ritchie, M. D., Hahn, L. W., Roodi, N., Bailey, L. R., Dupont, W. D., Parl, F. F., & Moore, J. H. (2001). Multifactor-dimensionality reduction reveals high-order interactions among estrogen-metabolism genes in sporadic breast cancer. *American Journal of Human Genetics*, 69, 138–147.

Schumacker, R. E. (2002). Latent variable interaction modeling. *Structural Equation Modeling: A Multidisciplinary Journal*, 9, 40–54.

REFERENCES

Aiken, L. S., & West, S. G. (1991). *Multiple regression: Testing and interpreting interactions*. Newbury Park, CA: Sage.

Algina, J., & Moulder, B. C. (2001). A note on estimating the Jöreskog–Yang model for latent variable interaction using LISREL 8.3. *Structural Equation Modeling,* 8(1), 40–52.

Bollen, K. A. (1989). *Structural equations with latent variables.* New York: John Wiley & Sons.

Bollen, K. A. (1995). Structural equation models that are nonlinear in latent variables: A least squares estimator. In P. M. Marsden (Ed.), *Sociological methodology.* Cambridge, MA: Blackwell.

Bollen, K. A. (1996). An alternative two stage least squares (2SLS) estimator for latent variable equations. *Psychometrika,* 61, 109–121.

Bollen, K. A., & Stine, R. A. (1993). Bootstrapping goodness-of-fit measures in structural equation models. In K. A. Bollen, & J. S. Long (Eds.), *Testing structural equation models* (pp. 66–110). Newbury Park, CA: Sage.

Cole, D. A., Maxwell, S. E., Arvey, R., & Salas, E. (1993). Multivariate group comparisons of variable systems: MANOVA and structural equation modeling. *Psychological Bulletin,* 114, 174–184.

du Toit, M., & du Toit, S. (2001). *Interactive LISREL: User's guide.* Lincolnwood, IL: Scientific Software International.

Etezadi-Amoli, J., & McDonald, R. P. (1983). A second generation nonlinear factor analysis. *Psychometrika,* 48, 315–342.

Higgins, L. F., & Judd, C. M. (1990). Estimation of non-linear models in the presence of measurement error. *Decision Sciences,* 21, 738–751.

Jaccard, J. & Turrisi, R. (2003). *Interaction effects in multiple regression* (2nd edn.). Thousand Oaks, CA: Sage Publications.

Jöreskog, K. G. (2000). *Latent variable scores and their uses.* Lincolnwood, IL: Scientific Software International.

Jöreskog, K. G., & Sörbom, D. (1993). *Bootstrapping and Monte Carlo experimenting with PRELIS2 and LISREL8.* Chicago: Scientific Software International.

Jöreskog, K. G., & Yang, F. (1996). Non-linear structural equation models: The Kenny–Judd model with interaction effects. In G. A. Marcoulides, & R. E. Schumacker (Eds.), *New developments and techniques in structural equation modeling* (pp. 57–88). Mahwah, NJ: Lawrence Erlbaum.

Jöreskog, K. G., Sörbom, D., Du Toit, S., & Du Toit, M. (2000). *LISREL8: New statistical features.* Lincolnwood, IL: Scientific Software International.

Kenny, D. A., & Judd, C. M. (1984). Estimating the non-linear and interactive effects of latent variables. *Psychological Bulletin,* 96, 201–210.

Klein, A., & Moosbrugger, H. (2000). Maximum likelihood estimation of latent interaction effects with the LMS method. *Psychometrika,* 65, 457–474.

Lunneborg, C. E. (1987). *Bootstrap applications for the behavioral sciences: Vol. 1.* Psychology Department, University of Washington, Seattle.

Mackenzie, S. B., & Spreng, R. A. (1992). How does motivation moderate the impact of central and peripheral processing on brand attitudes and intentions? *Journal of Consumer Research,* 18, 519–529.

Marsh, H. W., Wen, Z., & Hau, K.-T. (2004). Structural equation models of latent interactions: Evaluation of alternative estimation strategies and indicator construction. *Psychological Methods,* 9, 275–300.

McDonald, R. P. (1967). Nonlinear factor analysis. *Psychometric Monograph*, No. 15.

Mooney, C. Z., & Duval, R. D. (1993). *Bootstrapping: A nonparametric approach to statistical inference*. Sage University Series on Quantitative Applications in the Social Sciences, 07-097. Beverly Hills, CA: Sage.

Moulder, B. C., & Algina, J. (2002). Comparison of method for estimating and testing latent variable interactions. *Structural Equation Modeling*, 9(1), 1–19.

Newman, I., Marchant, G. J., & Ridenour, T. (1993, April). *Type VI errors in path analysis: Testing for interactions*. Paper presented at the annual meeting of the American Educational Research Association, Atlanta.

Ping, R. A., Jr. (1993). *Latent variable interaction and quadratic effect estimation: A suggested approach*. Technical Report. Dayton, OH: Wright State University.

Ping, R. A., Jr. (1994). Does satisfaction moderate the association between alternative attractiveness and exit intention in a marketing channel? *Journal of the Academy of Marketing Science*, 22(4), 364–371.

Ping, R. A., Jr. (1995). A parsimonious estimating technique for interaction and quadratic latent variables. *Journal of Marketing Research*, 32(3), 336–347.

Schumacker, R. E. (2002). Latent variable interaction modeling. *Structural Equation Modeling*, 9(1), 40–54.

Schumacker, R. E., & Marcoulides, G. A. (1998). *Interaction and nonlinear effects in structural equation modeling*. Mahwah, NJ: Lawrence Erlbaum.

Schumacker, R. E., & Rigdon, E. (1995, April). *Testing interaction effects in structural equation modeling*. Paper presented at the annual meeting of the American Educational Research Association, San Francisco.

Smith, K. W., & Sasaki, M. S. (1979). Decreasing multicollinearity: A method for models with multiplicative functions. *Sociological Methods and Research*, 8, 35–56.

Stine, R. (1990). An introduction to bootstrap methods: Examples and ideas. In J. Fox, & J. S. Long (Eds.), *Modern methods of data analysis* (pp. 325–373). Beverly Hills, CA: Sage.

Tabachnick, B. G., & Fidell, L. S. (2007). *Using multivariate statistics* (5th edn.). New York: Pearson Education.

Wong, S. K., & Long, J. S. (1987). *Parameterizing non-linear constraints in models with latent variables*. Unpublished manuscript, Indiana University, Department of Sociology, Bloomington, IN.

Yang-Wallentin, F., & Jöreskog, K. G. (2001). Robust standard errors and chi-squares in interaction models. In G. Marcoulides, & R. E. Schumacker (Eds.), *New developments and techniques in structural equation modeling* (pp. 159–171). Mahwah, NJ: Lawrence Erlbaum.

Chapter 16

REPORTING SEM RESEARCH

CHAPTER CONCEPTS

Checklist for structural equation modeling
Data preparation
Recommendations for modeling
Model specification
Model identification
Model estimation
Model testing
Model modification

OVERVIEW

Breckler (1990) reviewed the personality and social psychology research literature and found several shortcomings of structural equation modeling, namely
that model-fit indices can be identical for a potentially large number of models,
that assumptions of multivariate normality are required, that sample size affects
results, and that cross-validation of models was infrequently addressed or mentioned. Many of the studies only reported a single model-fit index. Breckler concluded that there was cause for concern in the reporting of structural equation
modeling results. Raykov, Tomer, and Nesselroade (1991) proposed guidelines for
reporting SEM results in the journal *Psychology and Aging*. Maxwell and Cole
(1995) offered some general tips for writing methodological articles, and Hoyle and
Panter (1995) published a chapter on reporting SEM research with an emphasis
on describing the results and what model-fit criteria to include.

The Publication Manual of the American Psychological Association (American
Psychological Association, 2001, pp. 161, 164–167, 185) specifically states that
researchers should include the means, standard deviations, and correlations of

the entire set of variables so that others can replicate and confirm the analysis, as well as provide example tables and figures for reporting structural equation modeling research. Unfortunately, the guidelines do not go far enough in outlining the basic information that should be included to afford an evaluation of the research study and some fundamental points that should be addressed when conducting SEM studies. A few other scholars have previously offered their advice, as follows.

Boomsma (2000) discussed how to write a research paper when structural equation models were used in empirical research and how to decide what information to report. His basic premise was that all information necessary for someone else to replicate the analysis should be reported. He provided recommendations along the lines of our basic steps in structural equation modeling, namely model specification, model identification, model estimation, model testing, and model modification. Boomsma found that many studies lacked a theoretical foundation for the theoretical model, gave a poor description of the model tested, provided no discussion of the psychometric properties of the variables and level of measurement, did not include sample data, and had a poor delineation or justification for the model modification process. He pointed out how difficult it can be to evaluate or judge the quality of published SEM research.

MacCallum and Austin (2000) provided an excellent survey of problems in applications of SEM. Thompson (2000) provided guidance for conducting structural equation modeling by citing key issues and including the following list of 10 commandments for good structural equation modeling behavior: (a) do not conclude that a model is the only model to fit the data, (b) cross-validate any modified model with split-sample data or new data, (c) test multiple rival models, (d) evaluate measurement models first, then structural models, (e) evaluate models by fit, theory, and practical concerns, (f) report multiple model-fit indices, (g) meet multivariate normality assumptions, (h) seek parsimonious models, (i) consider variable scale of measurement and distribution, and (j) do not use small samples.

McDonald and Ringo Ho (2002) examined 41 of 100 articles in 13 psychological journals from 1995 to 1997. They stated that SEM researchers should give a detailed justification of the SEM model tested along with alternative models, account for identification, address non-normality and missing data concerns, and include a complete set of parameter estimates with standard errors, correlation matrix (and perhaps residuals), and goodness-of-fit indices.

We further elaborate several key issues in SEM. First, in structural equation model analyses several different types of sample data matrices can be used (for example, variance–covariance matrix, asymptotic variance–covariance matrix, Pearson correlation matrix, or polyserial, polychoric, or tetrachoric matrices). As previously discussed, the type of matrix used depends on several factors such as type of variables and estimation method.

A second issue concerns model identification, that is, the number of distinct values in the sample variance–covariance matrix should equal or exceed the number of free parameters estimated in the model (degrees of freedom should not be negative for the model; the order condition) and the rank of the matrix should yield a non-zero determinant value (the rank condition). A researcher must also select from various parameter estimation techniques in model estimation, for example, unweighted least squares, maximum likelihood, or generalized least squares estimation under the assumption of multivariate normality, or asymptotically distribution-free estimation using Asymptotic Distribution Free (ADF) or CVM techniques when the multivariate normality assumption is not met. Obviously, many factors discussed in Chapters 2 and 3 affect multivariate normality.

A researcher should also be aware that equivalent models and alternative models may exist in an over-identified model (more distinct values in the matrix than free parameters estimated), and rarely are we able to perfectly reproduce the sample variance–covariance matrix, given the implied theoretical model. We use model-fit indices and specification searches to obtain an acceptable model to data fit, given alternative models. Model-fit statistics should guide our search for a better fitting model. Chapter 7 outlined different model-fit criteria depending on the focus of the research. Under some situations, for example, use of large sample sizes, the chi-square values will be inflated leading to statistical significance, and thus erroneously implying a poor data to model fit. A more appropriate use of the chi-square statistic in this situation would be to compare alternative models with the same sample data (nested models). The specification search process involves finding whether a variable should be added (parameter estimated) or deleted (parameter not estimated). A researcher, when modifying an initial model, should make one modification at a time, that is, add or delete one parameter estimate, and give a theoretical justification for the model change.

Ironically, structural equation modeling requires larger sample sizes as models become more complex or the researcher desires to conduct cross-validation with split samples. In traditional multivariate statistics the rule of thumb is 20 subjects per variable (20:1). The rules of thumb used in structural equation modeling varies from 100, 200, to 500 or more subjects per study, depending on model complexity and cross-validation requirements. Sample size and power are also important considerations in structural equation modeling (see Chapter 7). Finally, a two-step/four-step approach is important because if measurement models do not fit the observed variables, then relationships among the latent variables in structural models are not very meaningful.

We find the following suggestions to be valuable when publishing SEM research and hopefully journal editors will embrace the importance of this information when published.

1. Provide a review of literature that supports your theoretical model.
2. Provide the software program used along with the version.
3. Indicate the type of SEM model analysis (multi-level, structured means, etc.).
4. Include the correlation matrix, sample size, means, and standard deviations of variables.
5. Include a diagram of your theoretical model.
6. For interpretation of results, describe fit indices used and why; include power and sample size determination, and effect size measure.

Our suggestions are important because the SEM software, model, data, and program will be archived in the journal. The power, sample size, and effect size will permit future use in meta-analysis studies. Providing this research information will also permit future cross-cultural research, multi-sample or multi-group comparisons, replication, or validation by others in the research community because the analysis can be further examined.

We have made many of these same suggestions in the previous chapters, so our intentions in this chapter are to succinctly summarize guidelines and recommendations for SEM researchers. A brief summary checklist should help to remind us of the issues and analysis considerations a researcher makes.

CHECKLIST FOR STRUCTURAL EQUATION MODELING

Basic Issues

1. Is the sample size sufficient (power, effect size)?
2. Have you addressed missing data (MCAR, MAR, etc.)?
3. Have you addressed normality, outliers, linearity, restriction of range?
4. Have you checked the determinant, eigenvalues, and eigenvectors of matrix?
5. Are you using the correct covariance matrix?
6. Have you selected the correct estimation method?
7. Is the theoretical model identified ($df = 1$ or greater)?
8. Have you interpreted the direct, indirect, and total effects in the structural model?

Analysis Issues

1. Have you reported the correct fit indices?
2. Have you interpreted unstandardized and standardized significance of parameters?
3. Have you reported the factor variance for the set of indicator variables?
4. Have you justified any model modifications, for example, adding error covariances?
5. Have you cross-validated the model (assuming sufficient sample size)?
6. Have you diagrammed the model and/or provided estimates in the diagram?
7. Have you followed APA style guidelines?

We will follow through on this checklist of items as we discuss the modeling steps a researcher takes in conducting structural equation modeling.

DATA PREPARATION

A researcher should begin an SEM research study with a rationale and purpose for the study, followed by a sound theoretical foundation of the measurement model and the structural model. This includes a discussion of the latent variables and how they are defined in the measurement model. The hypothesis should involve the testing of the structural model and/or a difference between alternative models.

An applied SEM research study typically involves using sample data, in contrast to a methodological simulation study. The sample matrix should be described as to the type (augmented, asymptotic, covariance, or correlation), whether multivariate normality assumptions have been met, the scale of measurement for the observed variables, and should be related to an appropriate estimation technique, for example, maximum likelihood. Regression analysis, path analysis, factor analysis, and structural equation modeling all use data as input into a computer program. The SEM program should include the sample matrix, and for certain models, means and standard deviations of the observed variables.

Because sample variance–covariance matrices are used in SEM programs, it is important that a researcher knows the terminology, computation, and interpretation of determinant, eigenvalue, and eigenvector of the matrix. The determinant is the generalized variance of a matrix. The eigenvalue is the decomposition of the variance, while the eigenvectors provide the coefficients to produce the eigenvalues. There are several programs (LISREL9, R, SAS, SPSS) that compute these values. Basically, if the determinant is not positive and yields sufficient variance, then analysis is redundant. Similarly, if the eigenvalues are not positive, then we have issues with using the matrix to calculate parameter estimates.

A set of recommendations for data preparation is given in Table 16.1.

Table 16.1: Data Preparation Recommendations

1.	Have you adequately described the population from which the random sample data were drawn?
2.	Did you report the measurement level and psychometric properties (i.e., reliability and validity) of your variables?
3.	Did you report the descriptive statistics on your variables?
4.	Did you create a table with correlations, means, and standard deviations?
5.	Did you consider and treat any missing data (for example, which can result in data analysis issues)? What was the sample size both before and after treating the missing data?
6.	Did you consider and treat any outliers (for example, which can affect sample statistics)?
7.	Did you consider the range of values obtained for variables, as restricted range of one or more variables can reduce the magnitude of correlations?
8.	Did you consider and treat any non-normality of the data (for example, skewness and kurtosis, data transformations)?
9.	Did you consider and treat any multicollinearity among the variables?
10.	Did you consider whether variables are linearly related, which can reduce the magnitude of correlations?
11.	Did you resolve any correlation attenuation (for example, which can result in reduced magnitude of correlations and error messages)?
12.	Did you take the measurement scale of the variables into account when computing statistics such as means, standard deviations, and correlations?
13.	Did you specify the type of matrix used in the analysis (for example, covariance, correlation (Pearson, polychoric, polyserial), augmented moment, or asymptotic matrices)?
14.	When using the correlation matrix, did you include standard deviations of the variables in order to obtain correct estimates of standard errors for the parameter estimates?
15.	How can others access your data and SEM program (for example, appendix, website, email)?

MODEL SPECIFICATION

Model specification involves determining every relationship and parameter in the model that is of interest to the researcher. Moreover, the goal of the researcher is to determine, as well as possible, the theoretical model that generates a variance–covariance matrix similar to the sample variance–covariance matrix. If the theoretical model is misspecified, it could yield biased parameter estimates: parameter estimates that are different from what they are in the true population model, that is, specification error. We do not typically know the true population model, so bias in parameter estimates is generally attributed to specification error. The model should be developed from the available theory and research in the substantive area. This should be the main purpose of the literature review. A set of recommendations for model specification is given in Table 16.2.

Table 16.2: Model Specification Recommendations

1. Did you provide a rationale and purpose for your study, including why SEM rather than another statistical analysis approach was required?
2. Did you describe your latent variables, thus providing a substantive background to how they are measured?
3. Did you establish a sound theoretical basis for your measurement models and structural models?
4. Did you theoretically justify alternative models for comparison (for example, nested models)?
5. Did you use a reasonable sample size, thus sufficient power in testing your hypotheses?
6. Did you clearly state the hypotheses for testing the structural models?
7. Did you discuss the expected magnitude and direction of expected parameter estimates?
8. Did you include a figure or diagram of your measurement and structural models?

MODEL IDENTIFICATION

In structural equation modeling it is crucial that the researcher resolve the *identification problem* prior to the estimation of parameters in measurement models and/or structural models. In the identification problem, we ask the following question: On the basis of the sample data contained in the sample covariance matrix S, and the theoretical model implied by the population covariance matrix Σ, can a unique set of parameter estimates be found? A quick check on model identification is whether the degrees of freedom are equal to or greater than 1. Also, we check the value of the determinant of the matrix, eigenvalues, and eigenvectors. A set of recommendations for model identification is given in Table 16.3.

Table 16.3: Model Identification Recommendations

1. Did you specify the number of distinct values in your sample covariance matrix?
2. Did you indicate the number of free parameters to be estimated?
3. Did you inform the reader that the order and/or rank condition was satisfied?
4. Did you report the number of degrees of freedom and thereby the level of identification of the model?
5. How did you scale the latent variables (i.e., fix either one factor loading per latent variable or the latent variable variances to 1.0)?
6. Did you avoid non-recursive models until identification was assured?
7. Did you utilize parsimonious models to assist with identification?

MODEL ESTIMATION

In *model estimation* we need to decide which estimation technique to select for estimating the parameters in our measurement model and structural model, that is, our estimates of the population parameters from sample data. For example, we might choose the maximum likelihood estimation technique because we meet the multivariate normality assumption (acceptable skewness and kurtosis); there are no missing data; no outliers; and continuous variable data. If the observed variables are interval scaled and multivariate normal, then the ML estimates, standard errors, and chi-square test are appropriate. When we meet the multivariate normality assumption, and are using continuous variables that also meet the Pearson correlation assumptions, then the least squares and generalized least squares estimation methods provide accurate parameter estimates and standard errors. When we have messy data, then diagonally weighted least squares estimation is recommended using an asymptotic variance–covariance matrix. LISREL9 will automatically invoke this estimation strategy when inputting raw data; Mplus provides an estimation command to select other more advanced estimation methods, given categorical data.

Our experience is that model estimation often does not work because of messy data. In the earlier chapters, we outlined many of the factors that can affect parameter estimation in general, and structural equation modeling specifically. Missing data, outliers, multicollinearity, and non-normality of data distributions can seriously affect the estimation process and often result in fatal error messages pertaining to Heywood variables (variables with negative variance), non-positive definite matrices (determinant of matrix is zero), or failure to reach convergence (unable to compute a final set of parameter estimates). SEM is a correlation research method and all of the factors that affect correlation coefficients, the general linear model (regression, path, and factor models), and statistics in general are compounded in structural equation modeling. Researchers have the option of picking a different estimation method, using asymptotic distribution freer methods, and reporting a robust chi-square statistic. *Do not overlook the problems caused by messy data!* We have included some recommendations for model estimation in Table 16.4.

Table 16.4 Model Estimation Recommendations

1.	What is the ratio of chi-square to the degrees of freedom?
2.	What is the ratio of sample size to number of parameters?
3.	Did you consider tests of parameter estimates?
4.	Did you identify the estimation technique based on the type of data matrix?
5.	What estimation technique is appropriate for the distribution of the sample data (ML and GLS for multivariate normal data with small to moderate sample sizes; ADF or CVM for non-normal, asymptotic covariance data, and WLS for non-normal with large sample sizes)?
6.	Did you encounter Heywood cases (negative variance), multicollinearity, or non-positive definite matrices?
7.	Did you encounter and resolve any convergence problems or inadmissible solution problems by using start values, setting the admissibility check off, using a larger sample size, or using a different method of estimation?
8.	Which SEM program and version did you use?
9.	Did you report the R^2 values to indicate the fit of each separate equation?
10.	Do parameter estimates have the expected magnitude and direction?

MODEL TESTING

Having provided the SEM program and sample data along with the measurement and structural models, anyone can check our results and verify our findings. In interpreting our measurement model and structural model, we establish how well the data fit the models. In other words, we examine the extent to which the theoretical model is supported by the sample data. In model testing we consider model-fit indices for the fit of the entire model and examine the specific tests for the statistical significance of individual parameters in the model. A set of recommendations for model testing is provided in Table 16.5.

Table 16.5: Model Testing Recommendations

1. Did you report several model-fit indices (for example, for a single model: chi-square, df, GFI, NFI, RMSEA; for a nested model: LR test, CFI, AIC; for cross-validation indices: CVI, ECVI; and for parameter estimates, t values and standard errors)?
2. Did you specify separate measurement models and structural models?
3. Did you check for measurement invariance in the factor loadings prior to testing between-group parameter estimates in the structural model?
4. Did you provide a table of estimates, standard errors, statistical significance (possibly including effect sizes and confidence intervals)?

MODEL MODIFICATION

If the fit of an implied theoretical model is not acceptable, which is sometimes the case with an initial model, the next step is *model modification* and subsequent evaluation of the new, modified model. Most model modifications occur in the measurement model rather than the structural model. Model modification occurs more in the measurement model because that is where the main source of misspecification occurs and measurement issues are found.

After we are satisfied with our final best-fitting model, future research should undertake *model validation* by replicating the study, performing cross-validation (randomly splitting the sample and running the analysis on both sets of data), or bootstrapping the parameter estimates to determine the amount of bias. A set of recommendations for model modification is given in Table 16.6.

Table 16.6: Model Modification Recommendations

1. Did you compare alternative models or equivalent models?
2. Did you clearly indicate how you modified the initial model?
3. Did you provide a theoretical justification for the modified model?
4. Did you add or delete one parameter at a time? What parameters were trimmed?

Table 16.6: (*cont.*)

5. Did you provide parameter estimates and model-fit indices for both the initial model and the modified model?
6. Did you report statistical significance of free parameters, modification indices, and expected change statistics of fixed parameters, and residual information for all models?
7. How did you evaluate and select the best model?
8. Did you replicate your SEM model analysis using another sample of data?
9. Did you cross-validate your SEM model by splitting your original sample of data?
10. Did you use bootstrapping to determine the bias in your parameter estimates?

SUMMARY

In this chapter we explained that model fit is a subjective approach that requires substantive theory because there is no single best model (other models may be equally plausible given the sample data and/or equivalent models). In structural equation modeling the researcher follows the steps of model specification, identification, estimation, testing, and modification, so we advise the researcher to base measurement and structural models on *sound theory*, utilize the *two-step/four-step approach*, and establish measurement model fit and measurement invariance before *model testing* the latent variables in the structural model. We also recommend that theoretical models need to be *replicated, cross-validated*, and/or *bootstrapped* to determine the stability of the parameter estimates. Finally, we stated that researchers should include their SEM program, data, and/or path diagram of their model in any article. This permits a replication of the analysis and verification of the results. We do not advocate using specification searches to find the best-fitting model without having a theoretically justified reason for modifying the initial model. We further advocate using another sample of data to validate that the modified model is a meaningful and substantive theoretical structural model. We have therefore provided several recommendations for the five steps when doing structural equation modeling. These recommendations follow a logical progression from data preparation through model specification, identification, estimation, testing, and modification.

SUGGESTED READINGS

Boomsma, A. (2000). Reporting analyses of covariance structures. *Structural Equation Modeling*, 7, 461–483.
Cudeck, R., & MacCallum, R. C. (Eds). (2007). *Factor analysis at 100: Historical developments and future directions*. Mahwah, NJ: Erlbaum.

MacCallum, R. C., & Austin, J. T. (2000). Applications of structural equation modeling in psychological research. *Annual Review of Psychology*, 51, 201–226.

McDonald, R. P., & Ho, M. H. R. (2002). Principles and practice in reporting structural equation analyses. *Psychological Methods*, 7, 64–82.

Raykov, T., Tomer, A., & Nesselroade, J. R. (1991). Reporting structural equation modeling results in *Psychology and Aging*: Some proposed guidelines. *Psychology and Aging*, 6, 499–503.

Schreiber, J. B. (2008). Core reporting practices in structural equation modeling. *Research in Social and Administrative Pharmacy*, 4, 83–97.

Schreiber, J. B., Nora, A., Stage, F. K., Barlow, E. A., & King, J. (2006). Reporting structural equation modeling and confirmatory factor analysis results: A review. *Journal of Educational Research*, 99, 323–337.

Thompson, B. (2000). *Ten commandments of structural equation modeling*. In L. G. Grimm, & P. R. Yarnold (Eds.). *Reading and understanding more multivariate statistics* (pp. 261–283). Washington, DC: American Psychological Association.

REFERENCES

American Psychological Association. (2001). *Publication manual of the American Psychological Association* (5th edn.). Washington, DC: APA.

Boomsma, A. (2000). Reporting analyses of covariance structure. *Structural Equation Modeling*, 7, 461–483.

Breckler, S. J. (1990). Applications of covariance structure modeling in psychology: Cause for concern? *Psychological Bulletin*, 107, 260–273.

Hoyle, R. H., & Panter, A. T. (1995). Writing about structural equation models. In R. H. Hoyle (Ed.), *Structural equation modeling: Concepts, issues, and applications* (pp. 158–176). Thousand Oaks, CA: Sage.

MacCallum, R. C., & Austin, J. T. (2000). Applications of structural equation modeling in psychological research. *Annual Review of Psychology*, 51, 201–226.

Maxwell, S. E., & Cole, D. A. (1995). Tips for writing (and reading) methodological articles. *Psychological Bulletin*, 118, 193–198.

McDonald, R. P., & Ringo Ho, M. (2002). Principles and practice in reporting structural equation analyses. *Psychological Methods*, 7, 64–82.

Raykov, T., Tomer, A., & Nesselroade, J. R. (1991). Reporting structural equation modeling results in *Psychology and Aging*: Some proposed guidelines. *Psychology and Aging*, 6, 499–533.

Thompson, B. (2000). Ten commandments of structural equation modeling. In L. Grimm, & P. Yarnold (Eds.), *Reading and understanding more multivariate statistics* (pp. 261–284). Washington, DC: American Psychological Association.

Chapter 1

ANSWERS TO EXERCISES

1. Define the following terms:

   a. Latent variable: an unobserved variable that is not directly measured, but is computed using multiple observed variables.

   b. Observed variable: a raw score obtained from a test or measurement instrument on a trait of interest.

   c. Dependent variable: a variable that is measured and related to outcomes, performance, or criteria.

   d. Independent variable: a variable that defines mutually exclusive categories, for example, gender, region, or grade level), or is measured as a continuous variable, for example, test scores, and influences a dependent variable.

2. Explain the difference between a dependent latent variable and a dependent observed variable. A *dependent latent variable* is not directly measured, but is computed using multiple dependent observed variables. A *dependent observed variable* is a raw score obtained from a measurement instrument or assigned to a criterion variable.

3. Explain the difference between an independent latent variable and an independent observed variable. An *independent latent variable* is not directly measured, but is computed using multiple independent observed variables. An *independent observed variable* is a raw score obtained from a measurement instrument and used as a predictor variable.

4. List the four reasons why a researcher would conduct structural equation modeling.

   (i) Researchers are becoming more aware of the need to use multiple observed variables to better understand their area of scientific inquiry.

(ii) More recognition is given to the validity and reliability of observed scores from measurement instruments.

(iii) Structural equation modeling has improved recently, especially the ability to analyze more advanced statistical models.

(iv) SEM software programs have become increasingly user-friendly.

Chapter 2

ANSWERS TO EXERCISES

1. Define the following levels of measurement.

    a. Nominal: mutually exclusive groups or categories with number or percentage indicated.
    b. Ordinal: mutually exclusive groups or categories that are ordered with a ranking indicated.
    c. Interval: continuous data with arbitrary zero point, appropriately permitting a mean and a standard deviation.
    d. Ratio: continuous data with a true zero point, appropriately permitting a mean and a standard deviation.

2. Describe the different imputation methods.

    a. Mean substitution – compute mean on variable and use for missing value.
    b. Regression – compute predicted value for variable and use for missing value.
    c. EM – expected maximum algorithm computes and replaces missing value.
    d. Response pattern – use two variables with complete data to assign missing value.

3. Explain how each of the following affects statistics:

    a. Restriction of range: A set of scores that are restricted in range implies reduced variability. Variance and covariance are important in statistics, especially affecting correlation.
    b. Missing data: A set of scores with missing data can affect the estimate of the mean and standard deviation. It is important to determine whether the missing data are due to a data entry error, are missing at random, or are missing systematically due to some other variable (for example, gender).
    c. Outliers: A set of scores with an outlier (extreme score) can affect the estimate of the mean and standard deviation. It is important to determine

whether the outlier is an incorrect data value due to data entry error, represents another group of subjects not well sampled, or potentially requires the researcher to gather more data to fill in between the range of existing data.

d. Non-linearity: Researchers have generally analyzed relationships in data assuming linearity. Linearity is a requirement for the Pearson correlation coefficient. Consequently, a lack of linearity that is not included in the statistical model would yield misleading results.

e. Non-normality: Skewness, or lack of symmetry in the frequency distribution, and kurtosis, the departure from the peakedness of a normal distribution, affect inferential statistics, especially the mean, the standard deviation, and correlation coefficient estimates. Data transformations, especially a probit transformation, can help to yield a more normally distributed set of scores.

Chapter 3

ANSWERS TO EXERCISES

1. Partial and part correlations:

$$r_{12.3} = \frac{.6 - (.7)\,(.4)}{\sqrt{[1 - (.7)^2][1 - (.4)^2]}} = .49$$

$$r_{1(2.3)} = \frac{.6 - (.7)(.4)}{\sqrt{[1 - (.4)^2]}} = .35.$$

2. Bivariate = area $[a + c] = (.6)^2 = 36\%$
    Partial = area $[a / (a + e)] = (.49)^2 = 24\%$
    Part = area $a = (.35)^2 = 12\%$

3. A meaningful theoretical relationship should be plausible given that:

    a. Variables logically precede each other in time.
    b. Variables covary or correlate together as expected.
    c. Other influences or "causes" are controlled.
    d. Variables should be measured on at least an interval level.
    e. Changes in a preceding variable should affect variables that follow, either directly or indirectly.

4. The formula for calculating the Pearson correlation coefficient from the covariance and variances of variables is

$$r_{XY} = \frac{s^2_{XY}}{\sqrt{s^2_X * s^2_Y}}.$$

Therefore

$$r_{XY} = \frac{10.16}{\sqrt{(15.80)(11.02)}} = .77$$

$$r_{xz} = \frac{12.43}{\sqrt{(15.80)(15.37)}} = .80$$

$$r_{yz} = \frac{9.23}{\sqrt{(11.02)(15.37)}} = .71.$$

5. Enter the correlation matrix in a statistics package ($N = 209$). Calculate the determinant of the matrix, eigenvalues, and eigenvectors. For example, R uses det() and eigen() functions (see Rdet.r file). Report and interpret these values.

Correlation Matrix

Academic	1.00						
Athletic	.43	1.00					
Attract	.50	.48	1.00				
GPA	.49	.22	.32	1.00			
Height	.10	−.04	−.03	.18	1.00		
Weight	.04	.02	−.16	−.10	.34	1.00	
Rating	.09	.14	.43	.15	−.16	−.27	1.00
S. D.	.16	.07	.49	3.49	2.91	19.32	1.01
Means	.12	.05	.42	10.34	.00	94.13	2.65

For **SPSS** and **Excel**, the correlation, means, and standard deviations can be input as follows:

SPSS Matrix Input Example

	rowtype_	varname_	academic	athletic	attract	gpa	height	weight	rating
1	n		209.00	209.00	209.00	209.00	209.00	209.00	209.00
2	corr	acacemic	1.00	.	.	.	.	.	.
3	corr	athletic	.43	1.00	.	.	.	.	.
4	corr	attract	.50	.48	1.00	.	.	.	.
5	corr	GPA	.49	.22	.32	1.00	.	.	.
6	corr	height	.10	−.04	−.03	.18	1.00	.	.
7	corr	weight	.04	.02	−.16	−.10	.34	1.00	.
8	corr	rating	.09	.14	.43	.15	−.16	−.27	1.00
9	stddev		.16	.07	.49	3.49	2.91	19.32	1.01
10	mean		.12	.05	.42	10.34	.00	94.13	2.65

Microsoft Excel Matrix Input Example

In R we would create the following script file, *Rdet.r*, with the following lines of code.

```
# Chapter 16 - Determinant, eigenvalue, eigenvector of
   correlation matrix
# lower2full function in the lavaan package reads in lower
   triangle and yields square matrix
# rownames and colnames are added to correlation matrix
install.packages("lavaan")
library(lavaan)

corrmat = lower2full(c(1,.43,1,.50,.48,1,.49,.22,.32,1,.10,
-.04,-.03,.18,1,.04,.02,-.16,-.10,.34,1,.06,.14,.43,.15,-.16,
-.27,1))
rownames(corrmat) = colnames(corrmat) = c("Acad","Ath","Att",
"GPA","Hgt","Wgt","Rate")

corrmat
```

OUTPUT

```
> corrmat
      Acad  Ath   Att   GPA   Hgt   Wgt   Rate
Acad  1.00  0.43  0.50  0.49  0.10  0.04  0.06
Ath   0.43  1.00  0.48  0.22  -0.04 0.02  0.14
Att   0.50  0.48  1.00  0.32  -0.03 -0.16 0.43
```

```
GPA    0.49  0.22   0.32   1.00   0.18  -0.10   0.15
Hgt    0.10 -0.04  -0.03   0.18   1.00   0.34  -0.16
Wgt    0.04  0.02  -0.16  -0.10   0.34   1.00  -0.27
Rate   0.06  0.14   0.43   0.15  -0.16  -0.27   1.00
```

The determinant of correlation matrix is:

```
> det(corrmat)
[1] 0.2325538
```

The eigenvalues and eigenvectors of correlation matrix are computed as:

```
> eigen(corrmat)
$values
[1] 2.3733719 1.5517223 0.9131219 0.8136713 0.5464737 0.4724607
0.3291781
$vectors
             [,1]         [,2]         [,3]         [,4]        [,5]         [,6]        [,7]
[1,] -0.488817205  0.27283587 -0.05236451  0.30852466 -0.2411845 -0.46523933  0.5604206
[2,] -0.436431121  0.08846459 -0.57828348  0.02398608  0.3997531  0.53890233  0.1284386
[3,] -0.537900560 -0.08740090 -0.11488948 -0.24738157  0.1051064 -0.45958446 -0.6374650
[4,] -0.412294916  0.20113147  0.55404753  0.30742979 -0.2677718  0.48175979 -0.2903155
[5,]  0.003406014  0.59453619  0.40929453 -0.39386876  0.5462349 -0.07667097  0.1400339
[6,]  0.133066214  0.58343492 -0.36114984 -0.37365663 -0.5795855  0.10811691 -0.1556918
[7,] -0.305848472 -0.41925200  0.21148687 -0.67362795 -0.2549243  0.17960669  0.3673393
```

Summary

The determinant of the matrix is positive, which we desire. The eigenvalues for the seven variables are also positive. Recall that the determinant is equal to the product of the eigenvalues. The eigenvectors indicate a set of equation weights for each variable.

Chapter 4

ANSWERS TO EXERCISES

1. The following LISREL-SIMPLIS program is run to analyze the theoretical regression model for predicting gross national product (GNP) from knowledge of labor, capital, and time:

```
Regression of GNP
Observed variables: GNP LABOR CAPITAL TIME
Covariance Matrix:
 4256.530
 449.016 52.984
 1535.097 139.449 1114.447
 537.482 53.291 170.024 73.747
Sample Size: 23
Equation: GNP = LABOR CAPITAL TIME
Number of Decimals = 3
Path Diagram
End of Problem
```

2. Results indicated that GNP is significantly predicted ($R^2 = .997$). The three independent variables are statistically different from zero as indicated by $t > 1.96$ for each parameter estimate, so no model modification is necessary.

```
GNP  =  3.819*LABOR  +  0.322*CAPITAL  +  3.786*TIME,  Errorvar.=
   12.470, R² = 0.997
      (0.216)     (0.0305)      (0.186)    (4.046)
      17.698      10.540        20.351     3.082
```

The F test for statistical significance, effect size, and confidence interval for the theoretical regression model is

$$F = \frac{R^2 / p}{(1 - R^2)/n - p - 1} = \frac{.997/3}{(1 - .997)/19} = \frac{.3323}{.0001} = .3323$$

The effect size is $R^2 - [p / (n - 1)] = .997 - [3/22] = .86$. The 95% confidence interval is .991 to .998. The F test indicates a statistically significant R^2, the effect size is large, and the confidence interval for R^2 indicates that R^2 values will probably vary between .991 and .998 if the regression model is run with another sample of data.

## Chapter 5

ANSWER TO EXERCISE

LISREL PROGRAM

```
Achievement Path Model
 Observed Variables: Ach Inc Abl Asp
 Covariance matrix:
 25.500
 20.500 38.100
 22.480 24.200 42.750
 16.275 13.600 13.500 17.000
 Sample Size: 100
 Relationships
 Asp = Inc Abl
 Ach = Inc Abl Asp
 Print Residuals
 Options: ND = 3
 Path Diagram
 End of problem
```

PARTIAL LISREL OUTPUT

```
LISREL Estimates (Maximum Likelihood)
  Structural Equations

Ach = 0.645*Asp + 0.161*Inc + 0.231*Abl, Errorvar.= 6.507,
   R² = 0.745
      (0.0771)     (0.0557)     (0.0514)     (0.934)
        8.366        2.892        4.497        6.964
```

Asp = 0.244*Inc + 0.178*Abl, Errorvar.= 11.282, R^2 = 0.336
    (0.0690)   (0.0652)   (1.620)
     3.537   2.724   6.964

Covariance Matrix of Independent Variables

	Inc	Abl
Inc	38.100	
	(5.471)	
	6.964	
Abl	24.200	42.750
	(4.778)	(6.139)
	5.065	6.964

Goodness of Fit Statistics

Degrees of Freedom = 0
Minimum Fit Function Chi-Square = 0.00 (P = 1.000)
Normal Theory Weighted Least Squares Chi-Square = 0.00 (P = 1.000)

The Model is Saturated, the Fit is Perfect !

Chapter 6

ANSWERS TO EXERCISE

The following SIMPLIS program was written:

```
Confirmatory Factor Model Exercise Chapter 6
Observed Variables:
Academic Concept Aspire Degree Prestige Income
Correlation Matrix
1.000
0.487 1.000
0.236 0.206 1.000
0.242 0.179 0.253 1.000
0.163 0.090 0.125 0.481 1.000
0.064 0.040 0.025 0.106 0.136 1.000
Sample Size: 3094
Latent Variables: Motivate SES
Relationships:
  Academic - Aspire = Motivate
  Degree - Income = SES
Print Residuals
Number of Decimals = 3
Path Diagram
End of problem
```

Results overall suggest a less than acceptable fit:

```
Normal  Theory  Weighted  Least  Squares  Chi-Square  =  114.115
(P = 0.0)
Degrees of Freedom = 8
Root Mean Square Error of Approximation (RMSEA) = 0.0655
Standardized RMR = 0.0377
Goodness of Fit Index (GFI) = 0.988
```

Consequently, the model modification indices were examined. The largest decrease in chi-square results from adding an error covariance between Concept and Academic (boldfaced), thus allowing us to maintain a hypothesized two-factor model.

```
     The Modification Indices Suggest to Add the
     Path to   from        Decrease in Chi-Square   New Estimate
     Concept   SES                21.9                 -0.14
     Aspire    SES                78.0                  0.21
     Degree    Motivate           16.1                  0.31
     Prestige  Motivate           18.1                 -0.22

     The Modification Indices Suggest to Add an Error Covariance
     Between   and         Decrease in Chi-Square   New Estimate
     Concept   Academic           78.0                  0.63
     Aspire    Academic           21.9                 -0.12
     Degree    Aspire             75.3                  0.13
     Prestige  Concept             8.9                 -0.04
     Income    Degree             18.1                 -0.10
     Income    Prestige           16.1                  0.07
```

The following error covariance command line was added.

```
Let the errors Concept and Academic correlate
```

The results indicated further model modifications. The largest decrease in chi-square was determined to occur by adding an error covariance between Income and Prestige (boldfaced), thus maintaining our hypothesized two-factor confirmatory model.

```
     The Modification Indices Suggest to Add the
     Path to   from        Decrease in Chi-Square   New Estimate
     Degree    Motivate           20.3                  0.71
     Prestige  Motivate           18.4                 -0.39

     The Modification Indices Suggest to Add an Error Covariance
     Between   and         Decrease in Chi-Square   New Estimate
     Degree    Aspire             10.0                  0.09
     Prestige  Aspire              8.3                 -0.05
     Income    Degree             18.4                 -0.10
     Income    Prestige           20.3                  0.08
```

The following error covariance command line was added.

```
Let the errors Income and Prestige correlate
```

The final results indicated a more acceptable level of fit:

```
Normal  Theory  Weighted  Least  Squares  Chi-Square  =  14.519
   (P = 0.0243)
Degrees of Freedom = 6
Root Mean Square Error of Approximation (RMSEA) = 0.0214
Standardized RMR = 0.0123
```

Goodness of Fit Index (GFI) = 0.998. The final SIMPLIS program was:

```
Modified Confirmatory Factor Model - Exercise Chapter 6
Observed Variables:
Academic Concept Aspire Degree Prestige Income
Correlation Matrix
1.000
0.487 1.000
0.236 0.206 1.000
0.242 0.179 0.253 1.000
0.163 0.090 0.125 0.481 1.000
0.064 0.040 0.025 0.106 0.136 1.000
Sample Size: 3094
Latent Variables: Motivate SES
Relationships:
  Academic - Aspire = Motivate
  Degree - Income = SES
Let the errors Concept and Academic correlate
Let the errors Income and Prestige correlate
Print Residuals
Number of Decimals = 3
Path Diagram
End of problem
```

Chapter 7

ANSWERS TO EXERCISES

1. Define the following SEM modeling steps:

   a. Model specification: Developing a theoretical model to test based on all of the relevant theory, research, and information available.

   b. Model identification: Determining whether a unique set of parameter estimates can be computed given the sample data contained in the sample covariance matrix S and the theoretical model that produced the implied population covariance matrix Σ.

   c. Model estimation: Obtaining estimates for each of the parameters specified in the model that produced the implied population covariance matrix Σ. The intent is to obtain parameter estimates that yield a matrix Σ as close as possible to S, our sample covariance matrix of the observed or indicator variables. When elements in the matrix S minus the elements in the matrix Σ equal zero ($S - \Sigma = 0$), then $\chi^2 = 0$, indicating a perfect model fit to the data and all values in S are equal to values in Σ.

   d. Model testing: Determining how well the sample data fit the theoretical model. In other words, to what extent is the theoretical model supported by the obtained sample data? Global omnibus tests of the fit of the model as well as the fit of individual parameters in the model are available.

   e. Model modification: Changing the initial implied model and retesting the global fit and individual parameters in the new respecified model. To determine how to modify the model, there are a number of procedures available to guide the adding or dropping of paths in the model so that alternative models can be tested.

2. Define confirmatory models, alternative models, and model-generating approaches.
   In *confirmatory models*, a researcher can hypothesize a specific theoretical model, gather data, and then test whether the data fit the model.

In *alternative models*, a researcher specifies different models to see which model fits the sample data the best. A researcher usually conducts a chi-square difference test.

In *model generating*, a researcher specifies an initial model, then uses modification indices to modify and retest the model to obtain a better fit to the sample data.

3. Define model fit, model comparison, and model parsimony.

Model fit determines the degree to which the sample variance–covariance data fit the structural equation model.

Model comparison involves comparing an implied model with a null model (independence model). The null model could also be any model that establishes a baseline for expecting other alternative models to be different.

Model parsimony seeks the minimum number of estimated coefficients required to achieve a specific level of model fit. Basically, an over-identified model is compared with a restricted model.

4. Calculate the following fit indices for the model analysis in Figure 7.1:

$\text{GFI} = 1 - [\chi^2_{model} / \chi^2_{null}] = .97$

$\text{NFI} = (\chi^2_{null} - \chi^2_{model}) / \chi^2_{null} = .97$

$\text{RFI} = 1 - [(\chi^2_{model} / df_{model}) / (\chi^2_{null} / df_{null})] = .94$

$\text{IFI} = (\chi^2_{null} - \chi^2_{model}) / (\chi^2_{null} - df_{model}) = .98$

$\text{TLI} = [(\chi^2_{null} / df_{null}) - (\chi^2_{model} / df_{model})] / [(\chi^2_{null} / df_{null}) - 1] = .96$

$\text{CFI} = 1 - [(\chi^2_{model} - df_{model}) / (\chi^2_{null} - df_{null})] = .98$

$\text{Model AIC} = \chi^2_{model} + 2q = 50.41$

$\text{Null AIC} = \chi^2_{null} + 2q = 747.80$

$$RMSEA = \sqrt{[\chi^2_{Model} - df_{Model}]/[(N-1)df_{Model}]} = 0.083$$

5. How are modification indices in LISREL-SIMPLIS used?

Modification indices in LISREL-SIMPLIS indicate the amount of change in chi-square that would result if a path was added or dropped.

6. What steps should a researcher take in examining parameter estimates in a model?

A researcher should examine the sign of the parameter estimate, whether the value of the parameter estimate is within a reasonable range of values, and test the parameter for significance.

7. How should a researcher test for the difference between two alternative models?

A researcher computes a chi-square difference test (referred to as a likelihood ratio test) by subtracting the base model chi-square value from the constrained model chi-square value with one degree of freedom. When testing measurement invariance between groups in a specified model, the comparative fit index (CFI) and McDonald's non-centrality index (NCI) are recommended.

8. How are structural equation models affected by sample size and power considerations?

Several factors affect determining the appropriate sample size and power including model complexity, distribution of variables, missing data, reliability, and variance–covariance of variables. If variables are normally distributed with no missing data, sample sizes less than 500 should yield power = .80 and satisfy Hoelter's CN criterion. SAS, SPSS, G*Power 3 and other software programs can be used to determine power and sample size.

9. Describe the SEM modeling steps.

The four-step approach first uses exploratory factor analysis to establish a meaningful theoretical model. Next, one conducts a confirmatory factor analysis with a new sample of data. Then, one conducts a test of the structural equation model. Finally, a researcher tests planned hypotheses about free parameters in the model.

10. What new approaches are available to help researchers identify the best model?

The expected parameter change value has been added to LISREL output. Tabu and optimization algorithms have been proposed to identify the best model fit with the sample variance–covariance matrix.

11. Use G*Power 3 to calculate power for modified model with NCP = 6.3496 at $p = .05$, $p = .01$, $p = .001$ levels of significance. What happens to power when alpha increases?

Power decreases as alpha increases (power = .73, alpha = .05; power = .50, alpha = .01, and power = .24, alpha = .001).

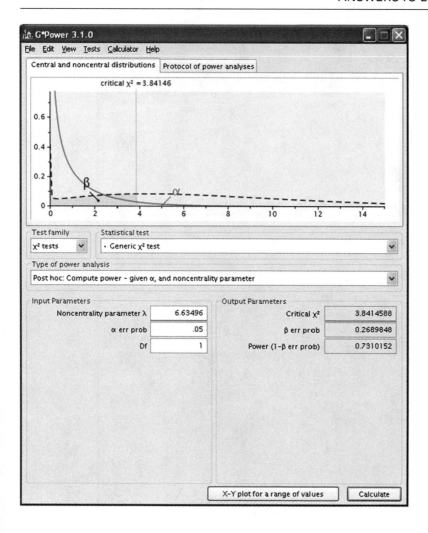

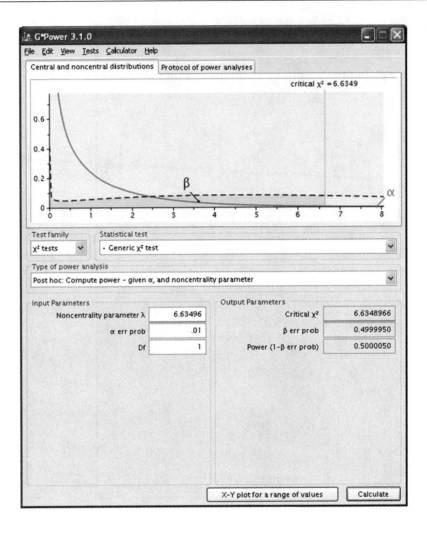

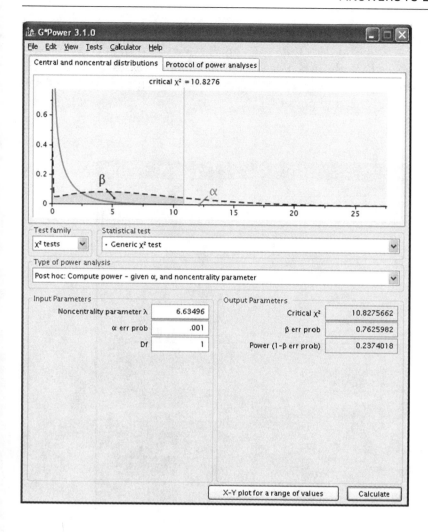

12. Use G*Power 3 to calculate power for a modified model with alpha = .05 and NCP = 6.3496 at df = 1, df = 2, and df = 3 levels of model complexity. What happens to power when degrees of freedom increases?
Power decreases as the degrees of freedom increases (power = .73, df = 1; power = .63, df = 2, and power = .56, df = 3).

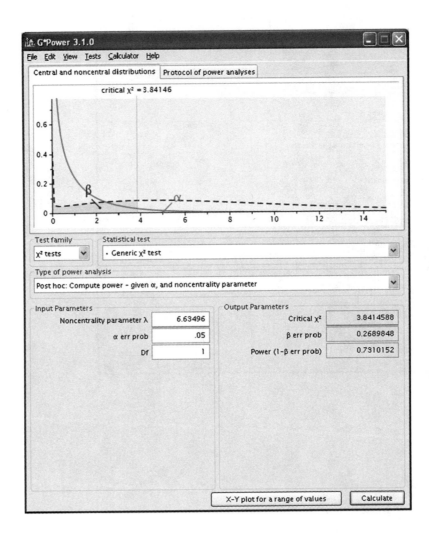

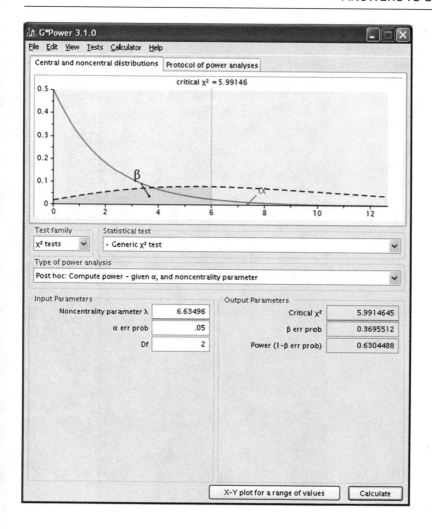

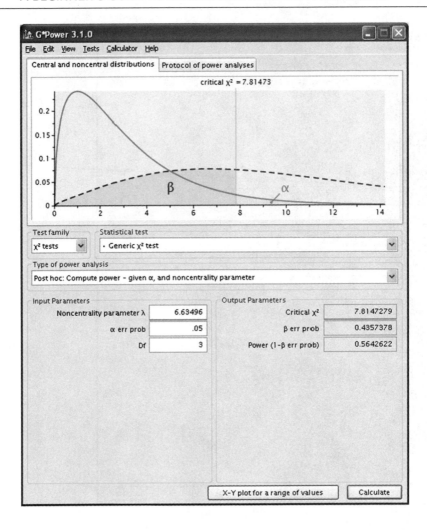

## Chapter 8

ANSWERS TO EXERCISES

1. MULTIPLE GROUP MODEL

You should run two separate SIMPLIS programs to obtain the results for the exercise. The first SIMPLIS program could test the null hypothesis by only having the EQUATION command in GROUP 1: Germany, thereby testing whether the path coefficients were the same (common model). *Note:* In a SIMPLIS multiple group model, all parameters in the model are constrained to be equal across groups unless specified to be free or equal.

Test of Same Path Model

The SIMPLIS program with only the EQUATION command in the GROUP 1: Germany is as follows:

```
Multiple Group Path Model Analysis - Same Model
Group 1: Germany
Observed Variables satis boss hrs type assist eval
Sample Size = 400
Means 1.12 2.42 10.34 4.00 54.13 12.65
Standard Deviation 1.25 2.50 3.94 2.91 9.32 2.01
Correlation Matrix
1.00
.55 1.00
.49 .42 1.00
.10 .35 .08 1.00
.04 .46 .18 .14 1.00
.01 .43 .05 .19 .17 1.00
```

Equation:

```
satis = boss hrs
boss = type assist eval satis
Let the errors of satis and boss correlate
Group 2: United States
Observed Variables satis boss hrs type assist eval
Sample Size = 400
Means: 1.10 2.44 8.65 5.00 61.91 12.59
Standard Deviations: 1.16 2.49 4.04 4.41 4.32 1.97
Correlation Matrix
1.00
.69 1.00
.48 .35 1.00
 .02 .24 .11 1.00
.11 .19 .16 .31 1.00
.10 .28 .13 .26 .18 1.00
Number of Decimals = 3
Path Diagram
End of Problem
```

Computer Output—Same Model

Germany

```
Group Goodness of Fit Statistics
Contribution to Chi-Square = 37.682
Percentage Contribution to Chi-Square = 50.821
Root Mean Square Residual (RMR) = 0.312
Standardized RMR = 0.0442
Goodness of Fit Index (GFI) = 0.973
```

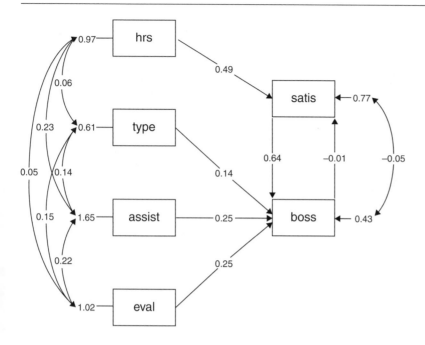

United States

```
Group Goodness of Fit Statistics
Contribution to Chi-Square = 36.464
Percentage Contribution to Chi-Square = 49.179
Root Mean Square Residual (RMR) = 0.332
Standardized RMR = 0.0420
Goodness of Fit Index (GFI) = 0.971
```

Global Goodness of Fit Statistics

```
Degrees of Freedom = 13
Minimum Fit Function Chi-Square = 74.146 (P = 0.00)
Normal Theory Weighted Least Squares Chi-Square = 69.772 (P = 0.00)
```

Note: Model modifications were indicated for both groups but not addressed here.

Test of Different Path Models

The second SIMPLIS program would put the EQUATION commands in both groups, GROUP 1: GERMANY and GROUP 2: UNITED STATES. The SIMPLIS program would look like this:

```
Multiple Group Path Model Analysis - Different Model
Group 1: Germany
Observed Variables satis boss hrs type assist eval
Sample Size = 400
Means 1.12 2.42 10.34 4.00 54.13 12.65
Standard Deviation 1.25 2.50 3.94 2.91 9.32 2.01
Correlation Matrix
1.00
.55 1.00
.49 .42 1.00
.10 .35 .08 1.00
.04 .46 .18 .14 1.00
.01 .43 .05 .19 .17 1.00
```

Equation:

```
satis = boss hrs
boss = type assist eval satis
Let the errors of satis and boss correlate
Group 2: United States
Observed Variables satis boss hrs type assist eval
Sample Size = 400
Means: 1.10 2.44 8.65 5.00 61.91 12.59
Standard Deviations: 1.16 2.49 4.04 4.41 4.32 1.97
Correlation Matrix
1.00
.69 1.00
.48 .35 1.00
.02 .24 .11 1.00
.11 .19 .16 .31 1.00
.10 .28 .13 .26 .18 1.00
Equation:
satis = boss hrs
boss = type assist eval satis
Let the errors of satis and boss correlate
Number of Decimals = 3
Path Diagram
End of Problem
```

Computer Output—Separate Path Models

Germany

```
Group Goodness of Fit Statistics
Contribution to Chi-Square = 5.415
Percentage Contribution to Chi-Square = 58.194
Root Mean Square Residual (RMR) = 0.166
 Standardized RMR = 0.0204
Goodness of Fit Index (GFI) = 0.995
```

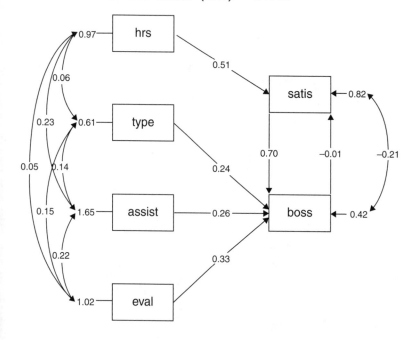

United States

```
Group Goodness of Fit Statistics
Contribution to Chi-Square = 3.890
Percentage Contribution to Chi-Square = 41.806
Root Mean Square Residual (RMR) = 0.113
Standardized RMR = 0.0208
Goodness of Fit Index (GFI) = 0.997
```

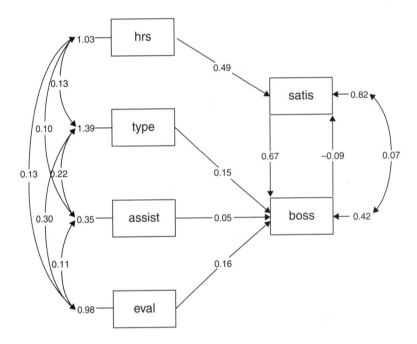

Global Goodness of Fit Statistics

```
Degrees of Freedom = 6
Minimum Fit Function Chi-Square = 9.306 (P = 0.157)
Normal Theory Weighted Least Squares Chi-Square = 9.199
(P = 0.163)
```

Summary

The chi-square values for the group path models under the assumption of a **similar** path model (same path coefficients) were not tenable (Germany: chi-square = 37.682 and United States: chi-square = 36.464) as indicated by the Global Chi-square value (chi-square = 74.146, degrees of freedom = 13, and $P = 0.000001$.

The chi-square values for the group path models under the assumption of **different** path models (different path coefficients) were tenable (Germany: chi-square = 5.415 and United States: chi-square = 3.89) as indicated by the Global Chi-square value (chi-square = 9.306, degrees of freedom = 6, and $P = 0.157$).

2. MULTIPLE SAMPLE MODEL

The two semesters of data only included correlation matrices (no means and standard deviations). *Note:* Although two samples are used, we still use the GROUP command. The SIMPLIS program is:

```
Multiple Sample - Predicting Clinical Competence in Nursing
Group 1: Semester 1
Observed variables comp effort learn
Sample size: 250
Correlation matrix
1.0
.28 1.0
.23 .25 1.0
Equation
comp = effort learn
Group 2: Semester 2
Observed variables comp effort learn
Sample size: 205
Correlation matrix
1.0
.21 1.0
.16 .15 1.0
Path Diagram
End of Problem
```

Computer Output—Multiple Sample Model

The regression model output indicated a non-significant chi-square (chi-square = .68, $df = 3$, $p = .88$), which implies that the two semesters of sample data had similar regression coefficients. We find that the regression coefficient of *effort* predicting *comp* is .22 compared to .24 and .19, respectively, in the two samples. We also find that the regression coefficient of *learn* predicting *comp* is .15 compared to .17 and .13, respectively, in the two samples. The correlation between effort and learn is .25 in the common regression model compared to .25 and .15, respectively, in the two samples of data. Finally, we see that the R-squared for the common regression model is .08 $(1 - R^2 = .92)$. The computer output indicated $R^2 = .10$ and .06, respectively, for the two regression equations from the two samples of data.

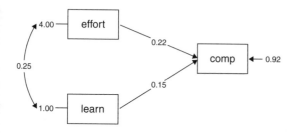

Chapter 9

ANSWERS TO EXERCISE

SECOND-ORDER FACTOR ANALYSIS

The psychological research literature suggests that drug use and depression are leading indicators of suicide among teenagers. The following SIMPLIS program was run to test a second-order factor model.

```
Second Order Factor Analysis
Observed Variables: drug1 drug2 drug3 drug4 depress1 depress2
depress3 depress4
Sample Size 200
Correlation Matrix
 1.000
 0.628 1.000
 0.623 0.646  1.000
 0.542 0.656  0.626  1.000
 0.496 0.557  0.579  0.640  1.000
 0.374 0.392  0.425  0.451  0.590  1.000
 0.406 0.439  0.446  0.444  0.668   .488  1.000
 0.489 0.510  0.522  0.467  0.643   .591   .612  1.000
Means   1.879  1.696  1.797  2.198  2.043  1.029  1.947  2.024
Standard Deviations 1.379 1.314 1.288 1.388 1.405 1.269   1.435
1.423
Latent Variables: drugs depress Suicide
Relationships
drug1 - drug4 = drugs
depress1 - depress4 = depress
drugs = Suicide
depress = Suicide
Set variance of drugs - Suicide to 1.0
Path Diagram
End of Problem
```

The second-order factor model with standardized coefficients had an acceptable fit (chi-square = 30.85, df = 19, p =.042) and is diagrammed as:

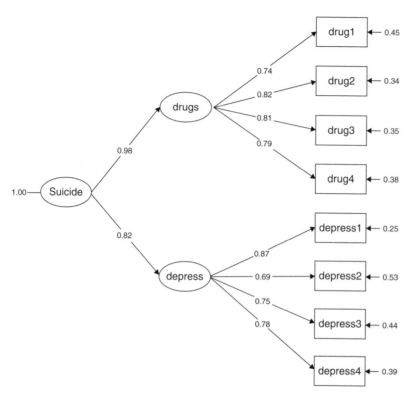

The structure coefficients indicate that the first factors are strong indicators of the second factor (suicide). Drug use (λ_{11} =.98; R^2 = 0.96) was a stronger indicator than depression (λ_{12} = .82; R^2 = 0.67) of suicide amongst teenagers.

Structural Equations

```
drugs = 0.98*Suicide, Errorvar.= 0.044, R² = 0.96
                        (0.17)
                        0.26
depress = 0.82*Suicide, Errorvar.= 0.33, R² = 0.67
          (0.12)                  (0.13)
          6.96                    2.51
```

Note: Missing t values and standard errors in SIMPLIS output are due to the software program.

Because the latent variables drugs and depress are indicators of the corresponding second-order latent variable suicide, LISREL by default fixes the loading of the first indicator to 1. Then, after convergence the value of 1 is rescaled using the estimated latent variable variance. Although the corresponding standard error estimate can be computed using the Delta method, LISREL does not compute and output the value. As a result, no standard error estimate and t value is written to the output file. The program with a raw data file should produce the standard errors and t values: another reason why using raw data can be beneficial to SEM analyses.

## Chapter 10

ANSWER TO EXERCISE

A sports physician was interested in studying heart rate and muscle fatigue of female soccer players. She collected data after three soccer games over a three-week period. A dynamic factor model was used to determine if heart rate and muscle fatigue were stable across time for the 150 female soccer players.

The SIMPLIS program to analyze the data is:

```
Dynamic Factor Model
Observed Variables: HR1 MF1 HR2 MF2 HR3 MF3
Covariance Matrix
   10.75
    7.00    9.34
    7.00    5.00   11.50
    5.03    5.00    7.49    9.96
    3.89    4.00    3.84    3.65    9.51
    2.90    2.00    2.15    2.88    3.55    5.50
Sample Size: 150
Latent Variables: Time1 Time2 Time3
Relationships:
 HR1 MF1 = Time1
 HR2 MF2 = Time2
 HR3 MF3 = Time3
 Time2 = Time1
 Time3 = Time2
Let the Errors of HR1 and HR2 correlate
Let the Errors of MF1 and MF2 correlate
Let the Errors of HR2 and HR3 correlate
Let the Errors of MF2 and MF3 correlate
Let the Errors of Time2 and Time3 correlate
```

```
Path Diagram
End of Problem
Let the Errors of HR2 and HR3 correlate
Let the Errors of MF2 and MF3 correlate
Let the Errors of Time2 and Time3 correlate
Path Diagram
End of Problem
```

The dynamic factor model had an acceptable model fit (chi-square = 5.54, $df = 2$, $p = .06$) and is diagrammed with standardized coefficients as:

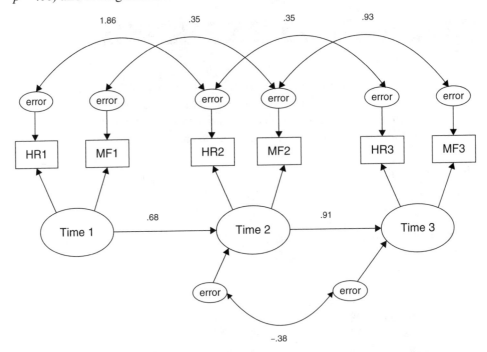

The structural equations indicate the prediction across the three time intervals for the latent variable, *time*. Time 1 predicted time 2; coefficient was statistically significant ($t = 6.30$) and R-squared = .47. Time 2 predicted time 3; coefficient was statistically significant ($t = 4.98$) and R-squared = .13.

Structural Equations

```
Time2 = 0.68*Time1, Errorvar.= 0.53, R² = 0.47
       (0.11)          (0.13)
        6.30            4.16
```

```
Time3 = 0.91*Time2, Errorvar.= 0.87, R² = 0.13
      (0.18)              (0.27)
       4.98               3.23
```

The dynamic factor model indicated that heart rate and muscle fatigue increase from time 1 to time 2 (coefficient = .68). Similarly, heart rate and muscle fatigue increase from time 2 to time 3 (coefficient = .91). The increased heart rate and muscle fatigue is of great concern to a sports physician. Proper rest and relaxation is therefore warranted to overcome the remaining effects of soccer games, especially on a weekly basis. The dynamic factor model clarifies how both heart rate and muscle fatigue increase over time.

Chapter 11

ANSWERS TO EXERCISE

The following SIMPLIS program would be created and run to determine the parameter estimates and model fit.

```
MIMIC Model of Job Satisfaction
Observed Variables peer self income shift age
Sample Size 530
Correlation Matrix
1.00
 .42 1.00
 .24 .35 1.00
 .13 .37 .25 1.00
 .33 .51 .66 .20 1.00
Latent Variable satisfac
Relationships
peer = satisfac
self = satisfac
satisfac = income shift age
Path Diagram
End of Problem
```

INITIAL MIMIC MODEL RESULTS

The MIMIC model results indicated an inadequate fit with chi-square = 6.81, $df = 2$, and $p = .033$. The measurement equations indicated that job satisfaction (satisfac) was adequately defined with self ratings being a better indicator of job satisfaction than peer ratings.

Measurement Equations

```
peer = 0.48*satisfac, Errorvar.= 0.77, R² = 0.23
                            (0.053)
                             14.49

self = 0.87*satisfac, Errorvar.= 0.25, R² = 0.75
       (0.11)                    (0.078)
        8.10                      3.16
```

The structural equation indicated that 45% of job satisfaction was predicted by knowledge of income, what shift a person worked, and their age. However, the coefficient for income was not statistically significant ($t = -.59$). Consequently, the model should be modified by dropping this variable and re-running the analysis.

Structural Equations

```
satisfac = -0.032*income + 0.31*shift + 0.56*age, Errorvar.=
    0.55, R² = 0.45
           (0.054)         (0.054)       (0.082)        (0.11)
           -0.59            5.71          6.77           5.14
```

MIMIC MODIFICATION

The MIMIC model modification resulted in little improvement with chi-square = 6.11, $df = 1$, and $p = .01$. The measurement equations were not very different. Other measures would help to define the latent variable, *job satisfaction*. The structural equation resulted in the same *R*-squared value, which indicates that income did not add to the prediction of *job satisfaction*. A parsimonious model was therefore achieved, but the 55% unexplained variance implies that other variables could be discovered to increase prediction.

Measurement Equations

```
peer = 0.49*satisfac, Errorvar.= 0.76 , R² = 0.24
                                 (0.053)
                                 14.48
self = 0.87*satisfac, Errorvar.= 0.25 , R² = 0.75
       (0.11)                    (0.078)
        8.12                      3.21
```

Structural Equations

```
satisfac = 0.31*shift + 0.54*age, Errorvar.= 0.55, R² = 0.45
           (0.053)      (0.073)               (0.11)
            5.72         7.39                  5.14
```

Chapter 12

ANSWERS TO EXERCISES

1. MIXED VARIABLE MODEL—DIFFERENT VARIABLE TYPES

The following LISREL program will test the Mixed Variable Model as specified.

```
Mixed Variable Model
Observed Variables: Age Gender Degree Region Hours Income
Correlation Matrix
 1.000
 0.487  1.000
 0.236  0.206  1.000
 0.242  0.179  0.253  1.000
 0.163  0.090  0.125  0.481  1.000
 0.064  0.040  0.025  0.106  0.136  1.000
Means 15.00  10.000  10.000 10.000  7.000  10.000
Standard Deviations  10.615  10.000 8.000  10.000 15.701 10.000
Sample Size 600
Latent Variables Person Earning
Relationships
Age Gender Degree = Person
Region Hours Income = Earning
Earning = Person
Number of Decimals = 3
Path Diagram
End of Problem
```

The Mixture Model with standardized parameter solution is:

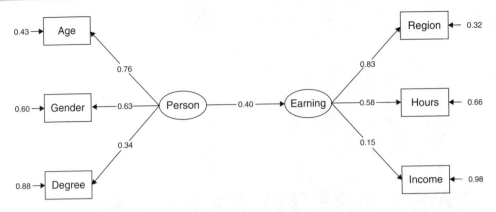

The model fit is not adequate (chi-square = 22.10, *df* = 8, *p* = .0047), so model modifications are indicated in the computer output.

Model Modification

The modification indices suggest the following:

```
The Modification Indices Suggest to Add an Error Covariance
Between  and      Decrease in Chi-Square    New Estimate
Gender  Age           15.1                   66.45
Degree  Region        14.6                   10.48
```

Gender and Age are a sensible modification because Degree and Region define different latent variables. Add the following line in the LISREL program after the last command:

```
Let error covariance between Gender and Age correlate
```

The LISREL output now produces a mixed variable model with acceptable model fit (chi-square = 6.898, *df* = 7, *P* = 0.440).

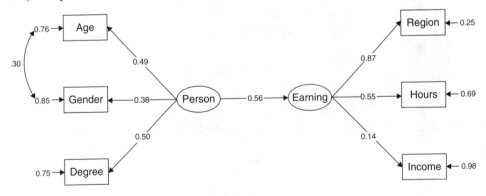

Measurement Equations

Region = 8.680*Earning, Errorvar.= 24.665, R² = 0.753
                              (12.232)
                               2.016

Hours = 8.699*Earning, Errorvar.= 170.855, R² = 0.307
        (1.486)   (15.724)
         5.855   10.866

Income = 1.385*Earning, Errorvar.= 98.081, R² = 0.0192
         (0.496)   (5.702)
          2.795   17.202

Age = 5.151*Person, Errorvar.= 86.149, R² = 0.235
      (0.695)   (7.642)
       7.413   11.273

Gender = 3.841*Person, Errorvar.= 85.245, R² = 0.148
         (0.652)   (6.340)
          5.895   13.445

Degree = 4.004*Person, Errorvar.= 47.972, R² = 0.250
         (0.516)   (4.332)
          7.758   11.073

Error Covariance for Gender and Age = 31.910
                                      (5.651)
                                       5.647

Structural Equations

Earning = 0.565*Person, Errorvar.= 0.681, R² = 0.319
          (0.0740)   (0.171)
           7.631   3.973

Summary

The mixed variable model is a special type of SEM model that uses continuous and ordinal variables in polyserial or polychoric matrices. Personal characteristics

(Person) predicted Earning Power (Earning), i.e., R-squared $= .319$. The mixed variable model required modification, namely correlate error covariance between Age and Gender, to achieve acceptable model fit criteria. This made theoretical sense because more males than females were in the workforce data set that was used in the model.

2. MIXTURE MODEL—BINARY LATENT CLASS

We used the data set from www.math.smith.edu/r/data/help.csv to conduct an LCA using the following 4 variables: homeless, cesdcut, satreat, and linkstat, with $n = 431$ data points ($n = 453$ with 22 missing cases). The binary variables are coded as follows: homeless (1 = yes; 0 = no); cesdcut (1 = high > 20; 0 = low < 20); satreat (1 = substance abuse treatment; 0 = no treatment); linkstat (1 = linked primary care; 0 = no primary care). For convenience, the reduced data set, *newds.txt*, is provided on the book website for this exercise. We hypothesized that there exist two latent classes: homeless with substance abuse, and homeless without substance abuse. Do the results support two latent classes?

The Mplus program is:

```
TITLE:LCA with binary latent class indicators
DATA:FILE IS newds.txt;
VARIABLE: NAMES ARE homeless cesdcut satreat linkstat;
    CLASSES = c (2);
    CATEGORICAL = homeless cesdcut satreat linkstat;
ANALYSIS:TYPE = MIXTURE;
OUTPUT:TECH11 TECH14;
```

Univariate percentages seem to indicate different percentages for two categories as follows:

```
UNIVARIATE PROPORTIONS AND COUNTS FOR CATEGORICAL VARIABLES
    HOMELESS
        Category 1    0.534    230.000
        Category 2    0.466    201.000
    CESDCUT
        Category 1    0.176     76.000
        Category 2    0.824    355.000
```

```
SATREAT
   Category 1    0.705    304.000
   Category 2    0.295    127.000
LINKSTAT
   Category 1    0.622    268.000
   Category 2    0.378    163.000
```

However, the Pearson and Likelihood ratio chi-square tests of model fit were significant. Yet the estimated model showed a percentage difference in group membership: Group 1 = 361 (84%); Group 2 = 70 (16%).

```
Chi-Square Test of Model Fit for the Binary and Ordered
Categorical
(Ordinal) Outcomes
      Pearson Chi-Square
      Value                                  17.107
      Degrees of Freedom                        6
      P-Value                                0.0089
      Likelihood Ratio Chi-Square
      Value                                  18.136
      Degrees of Freedom                        6
      P-Value                                0.0059
FINAL CLASS COUNTS AND PROPORTIONS FOR THE LATENT CLASSES
BASED ON THE ESTIMATED MODEL
      Latent
      Classes
        1        360.91675      0.83739
        2         70.08325      0.16261
```

The two reported tests for the H_o versus the H_a alternative hypothesis of two latent classes was not significant. Therefore, we would conclude that two latent classes are **not** present in the distribution of data.

```
VUONG-LO-MENDELL-RUBIN LIKELIHOOD RATIO TEST FOR 1 (H0) VERSUS
2 CLASSES
      H0 Loglikelihood Value                 -1045.656
      2 Times the Loglikelihood Difference      9.431
      Difference in the Number of Parameters        5
      Mean                                     76.118
      Standard Deviation                      104.227
      P-Value                                  0.7615
```

```
PARAMETRIC BOOTSTRAPPED LIKELIHOOD RATIO TEST FOR 1 (H0) VERSUS
2 CLASSES
        H0 Loglikelihood Value                      -1045.656
        2 Times the Loglikelihood Difference            9.431
        Difference in the Number of Parameters              5
        Approximate P-Value                            0.2000
        Successful Bootstrap Draws                         25
```

Chapter 13

ANSWERS TO EXERCISES

1. MULTI-LEVEL MODEL—OBSERVED VARIABLES

The multi-level analysis of data in the PRELIS system file, *income.lsf*, was used with the pull-down multi-level menu to create and run three different PRELIS programs. Results are summarized in a table with the intraclass correlation (hand computed) for comparative purposes.

Model 1 is the baseline model (constant), followed by the added effects of gender, and the added effects of marital status (marital). The three different PRELIS programs should look as follows:

Model 1 (intercept only)

```
OPTIONS OLS=YES CONVERGE=0.001000 MAXITER=10 OUTPUT=STANDARD;
 TITLE=income decomposition;
 SY='C:\LISREL 8.8 Student Examples\MLEVELEX\INCOME.PSF';
 ID3=region;
 ID2=state;
 RESPONSE=income;
 FIXED=constant;
 RANDOM2=constant;
 RANDOM3=constant;
```

Model 2 (intercept + gender)

```
OPTIONS OLS=YES CONVERGE=0.001000 MAXITER=10 OUTPUT=STANDARD;
 TITLE=income decomposition;
 SY='C:\LISREL 8.8 Student Examples\MLEVELEX\INCOME.PSF';
```

```
ID3=region;
ID2=state;
RESPONSE=income;
FIXED=constant gender;
RANDOM2=constant;
RANDOM3=constant;
```

Model 3 (intercept + gender + marital)

```
OPTIONS OLS=YES CONVERGE=0.001000 MAXITER=10 OUTPUT=STANDARD;
TITLE=income decomposition;
SY='C:\LISREL 8.8 Student Examples\MLEVELEX\INCOME.PSF';
ID3=region;
ID2=state;
RESPONSE=income;
FIXED=constant gender marital;
RANDOM2=constant;
RANDOM3=constant;
```

The PRELIS program results for the three analyses are summarized in Table 1. The baseline model (intercept only) provides the initial breakdown of level 3 and level 2 error variance. The multi-level model for the added effect of *gender* is run next. The chi-square difference between Model 1 and Model 2 yields chi-square = 5.40, which is statistically significant at the .05 level of significance. *Gender* therefore does help explain variability in *income*. Finally, *marital* is added to the multi-level

Table 1: Summary Results for Multi-level Analysis of Income

Multi-level Model Fixed Factors	Model 1 Constant	Model 2 Constant + Gender	Model 3 Constant + Gender + Marital
Intercept Only (B_0)	10.096 (.099)	10.37 (.15)	10.24 (.19)
Gender (B_1)		−0.42 (.16)	−0.43 (.16)
Marital (B_2)			.19 (.17)
Level 2 error variance (e_{ij})	.37	.31	.30
Level 3 error variance (u_{ij})	.02	.05	.06
ICC	.051 (5%)	.138 (14%)	.166 (17%)
Deviance (−2LL)	11144.29	11138.89	11137.71
df	3	4	5
Chi-square Difference (*df* = 1)		5.40	1.18

Note: $\chi^2 = 3.84$, *df* = 1, p = .05.

model, which yields a chi-square difference between Model 2 and Model 3 of chi-square = 1.18. The chi-square difference value is not statistically significant; therefore marital status does not add any additional significant explanation of variability in *income*.

Note:

$$ICC_1 = \frac{\Phi_3}{\Phi_3 + \Phi_2} = \frac{Tau - Hat(Level - 3)}{Tau - Hat(Level - 3) + Tau - Hat(Level - 2)} = \frac{.02}{.02 + .37} = .051$$

2. MULTI-LEVEL MODEL—LATENT VARIABLE

The multi-level model with two covariates (x1, x2) is given in Figure 13.2.

Figure 13.2: Multi-level CFA Model—Latent Variable

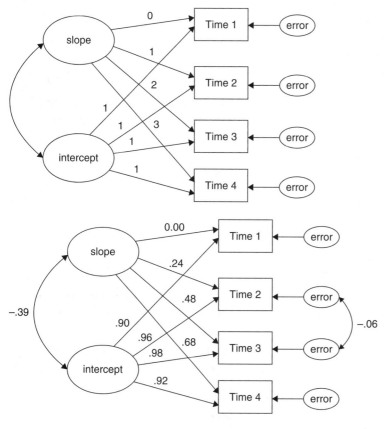

The Mplus program for Figure 13.2 with two covariates is given as:

```
TITLE:  Two-level CFA with continuous indicators, random inter-
cept, and covariates
DATA:  FILE IS ex9.6.dat;
VARIABLE:  NAMES ARE y1-y4 x1 x2 w clus;
Usevariables are ALL;
  WITHIN = x1 x2;
  BETWEEN = w;
  CLUSTER = clus;
ANALYSIS:TYPE = TWOLEVEL;
MODEL:
  %WITHIN%
  fw BY y1-y4;
  fw ON x1 x2;
  %BETWEEN%
  fb BY y1-y4;
  y1-y4@0;
  fb ON w;
OUTPUT: SAMPSTAT STANDARDIZED;
```

The questions are:

a. Are the intraclass correlation coefficients $> .10$ to indicate that multi-level modeling of variance across the 10 groups in the example data set? *Yes.*

The results indicate 110 clusters with average cluster size of 9. The intraclass correlations are all $> .10$, so multi-level analysis is appropriate.

```
SUMMARY OF DATA
      Number of clusters                        110
      Average cluster size                    9.091
      Estimated Intraclass Correlations for the Y Variables
```

Variable	Intraclass Correlation	Variable	Intraclass Correlation	Variable	Intraclass Correlation	Variable	Intraclass Correlation
Y1	0.125	Y2	0.121	Y3	0.106	Y4	0.115

b. What is the chi-square model-fit value? Is the chi-square test of model fit non-significant, which indicates a good data to model fit? *Yes.*

```
Chi-Square Test of Model Fit
      Value                     5.368
      Degrees of Freedom          17
      P-Value                  0.9965
```

c. Are the X1 and X2 within-model covariates statistically significant? *Yes.*

```
FW  ON
                                              Two-Tailed
            Estimate     S.E.     Est./S.E.   P-Value
    X1       0.973      0.048      20.256       0.000
    X2       0.510      0.039      13.017       0.000
```

d. Are the Y1–Y4 factor loadings significantly different in the between-factor model? *Yes.*

```
Between Level
FB     BY
                                              Two-Tailed
            Estimate     S.E.     Est./S.E.   P-Value
    Y1       0.677      0.069       9.768       0.000
    Y2       0.649      0.064      10.079       0.000
    Y3       0.625      0.072       8.634       0.000
    Y4       0.642      0.069       9.350       0.000
```

e. Are the Y1–Y4 intercepts significantly different in the between-factor model? *No.*

```
Between Level
Intercepts
    Y1      -0.083      0.072      -1.143       0.253
    Y2      -0.077      0.073      -1.058       0.290
    Y3      -0.045      0.069      -0.647       0.518
    Y4      -0.030      0.072      -0.413       0.679
```

Chapter 14

ANSWERS TO EXERCISES

1. LGM—INTERCEPT AND SLOPE

News and radio stations in the City of Dallas have apparently convinced the public that a massive crime wave has occurred during the past 4 years, from 2002 to 2005. A criminologist gathered the crime rate data, but needs your help to run a latent growth curve model to test whether a linear trend in crime rates exists for the City of Dallas.

The LISREL-SIMPLIS latent growth curve model is:

```
Latent Growth Model
Observed variables: time1 time2 time3 time4
Sample Size 400
Correlation Matrix
1.000
 .799 1.000
 .690  .715 1.000
 .605  .713  .800 1.000
Means 5.417 5.519 5.715 5.83
Standard Deviations .782 .755 .700 .780
Latent Variables: slope intercept
Relationships:
time1 =  CONST + 0 * slope + 1 * intercept
time2 =  CONST + 1 * slope + 1 * intercept
time3 =  CONST + 2 * slope + 1 * intercept
time4 =  CONST + 3 * slope + 1 * intercept
Let slope and intercept correlate
Path Diagram
End of Problem
```

The latent growth curve model is diagrammed as:

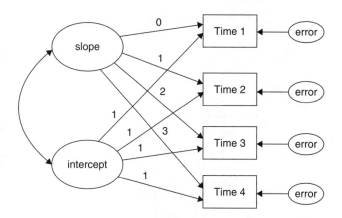

The LISREL-SIMPLIS results indicated a poor model fit (chi-square = 15.90, $df = 3$, $p = 0.001$). The group means, however, suggest an increase in crime rate from 2002 to 2005, so model modification was conducted:

```
Means
     time1        time2        time3        time4
   --------     --------     --------     --------
     5.42         5.52         5.71         5.83
```

Modification indices were indicated that recommended correlating the error covariance between Time 2 and Time 3. The crime rate had the greatest increase in this time period so possibly measurement disturbance was present.

Model Modification

The LISREL-SIMPLIS program was re-run with the following added command:

```
Let error covariance between time2 and time3 correlate
```

After modification the latent growth curve model had a more acceptable model fit (chi-square = 6.67, $df = 2$, and $p = .036$).

The final latent growth curve model with standardized coefficients is diagrammed as:

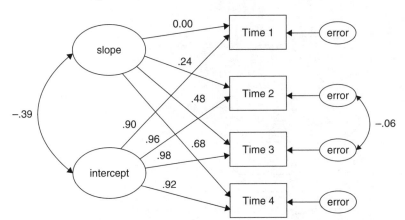

The individual slopes increased over time:

```
GroupSlope
Time 1.00
Time 2.24
Time 3.48
Time 4.68
```

The intercepts increased over the first 3 years, then decreased:

```
GroupIntercept
Time 1.90
Time 2.96
Time 3.98
Time 4.92
```

The negative correlation between slope and intercept occurred due to intercept dropping in value during year 2005 ($r = -.39$), so it does not clearly reflect the visual increase in both slope and intercept values over time. A practical interpretation is that crime rates have increased over time, which would be seen when plotting the mean crime rates over time; however, probably not a linear trend in means because of the drop from Time 3 to Time 4 intercepts. Crime rates in 2002 and 2003 were similar as well as crime rates in 2004 and 2005. LGM permitted the error covariance, which is not permitted in traditional repeated measures ANOVA.

2. LGM—GROUP DIFFERENCE

The data set *region.txt* contains the number of tornado sightings in Oklahoma and Virginia for 3 months, sighted by 10 individuals in each state. Region is dummy coded (1 = Oklahoma; 0 = Virginia). Run the R program to determine if the intercepts and/or slopes differ for each region. Alternatively, run the LGM in a different SEM software program. Report the values.

```
# Read in Data set
LGdata = read.table(file="c:/region.txt",header=TRUE,sep=",")
head(LGdata)
# lavaan SEM package
install.packages("lavaan")
library(lavaan)
# LG model
sighting = '
# intercept
i = ~ 1*y1 + 1*y2 + 1*y3
# slope
s = ~ 0*y1 + 1*y2 + 2*y3
# regression
i + s = ~ region
'

# Run LGM
model = growth(sighting,data=LGdata)
summary(model)
```

A brief display of the raw data is given as:

```
> head(LGdata)
id y1 y2 y3 region
1   1  1  3  5     1
2   2  1  4  6     1
3   3  2  5  7     1
4   4  2  4  5     1
5   5  3  5  6     1
6   6  3  6  7     1
```

The data fit the LGM linear model ($\chi^2 = 3.982$, $df = 3$, $p = .263$). Results indicated that the number of tornado sightings (intercepts) in Oklahoma and Virginia differed by region ($z = 4.281$, $p < .0001$). There was no significant rate of change difference (slope) across the time periods for the two regions ($z = .713$, $p = .476$). This is indicated in the output for the intercept (i) and slope (s) equations:

```
Estimate   Std.err    Z-value    P(>|z|)
i =~
region     0.225   0.052   4.281   0.000
s =~
region     0.055   0.077   0.713   0.476
```

The R output was given as:

```
> summary(model)
lavaan (0.5-17) converged normally after 75 iterations
 Number of observations                         20
 Estimator                                       ML
 Minimum Function Test Statistic              3.982
 Degrees of freedom                               3
 P-value (Chi-square)                         0.263
Parameter estimates:
 Information                               Expected
 Standard Errors                           Standard
                  Estimate  Std.err Z-value P(>|z|)
Latent variables:
 i =~
   y1             1.000
   y2             1.000
   y3             1.000
 s =~
   y1             0.000
   y2             1.000
   y3             2.000
 i =~
   region         0.225    0.052   4.281   0.000
 s =~
   region         0.055    0.077   0.713   0.476
Covariances:
 i ~~
 s               1.455    0.504   2.884   0.004
Intercepts:
 y1             0.000
 y2             0.000
 y3             0.000
 region         0.000
 i             2.126    0.277   7.682   0.000
 s             1.268    0.191   6.650   0.000
```

```
Variances:
  y1              1.758    0.674
  y2             -0.254    0.233
  y3              2.373    0.920
  region          0.127    0.046
  i               0.784    0.536
  s              -0.228    0.400
```

## Chapter 15

ANSWERS TO EXERCISES

1. KENNY–JUDD PRODUCT INDICANT MODEL

Figure 15.5 shows three observed indicator variables (V11, V12, V13) and an interaction product indicator variable (V17_21) for *Self*. The CFA model is testing whether the interaction of Age and Stress affects Sense of Self (Self). The variables used in the data were: V11 = Self-Esteem; V12 = Locus of Control: V13 = Marital Satisfaction; V17 = Age; V21 = Life Change (Stress). The interaction variable was created as: V17_21 = V17 * V21.

The example is a subset of an analysis by Tabachnick and Fidell (2007, 5th edition).

The data were downloaded from the Internet, and missing values (0) were inserted, *N* = 459. Data: media.pearsoncmg.com/ab/ab_tabachnick_multistats_6/datafiles/ASCII/hlthsem.dat. The subset of data used in the exercise is available on the book website in the file, *health.txt*.

Write an SEM program to test the statistical significance of the parameter estimates (paths) in the model.

Figure 15.5: Self Latent Variable with Interaction Variable

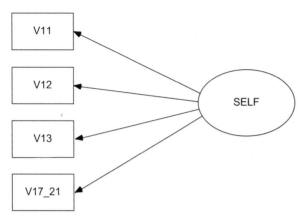

A SIMPLIS program was written using the raw data (*health.txt*) imported and saved as a LISREL system file, *final.lsf*.

```
Kenny Judd Model Chap 15.5 Exercise 1
Raw Data from File final.lsf
Observed variables: V11 V12 V13 V17 V21 V17_21
Latent variables Self
Relationships
V11 V12 V13 V17_21 = Self
Path Diagram
End of Problem
```

a. Did the data fit the model? *Yes*. ($\chi^2 = 1.3$, $df = 2$, $p = .52$)

```
Goodness of Fit Statistics
 Degrees of Freedom for (C1)-(C2)                              2
 Maximum Likelihood Ratio Chi-Square (C1)    1.308 (P = 0.5199)
 Browne's (1984) ADF Chi-Square (C2_NT)      1.301 (P = 0.5217)
 Estimated Non-centrality Parameter (NCP)                    0.0
 90 Percent Confidence Interval for NCP             (0.0; 6.122)
```

b. Were all parameter estimates statistically significant? *No*. The interaction variable was not statistically significant (V17_21 = –.669; $z = -1.806$, $p = .07$).

```
Measurement Equations

       V11 = 2.877*Self, Errorvar.= 7.333, R² = 0.530
  Standerr   (0.369)              (1.978)
  Z-values    7.803                3.708
  P-values    0.000                0.000
```

```
        V12 = 0.592*Self, Errorvar.= 1.252, R² = 0.219
Standerr   (0.0887)                  (0.116)
Z-values    6.672                    10.780
P-values    0.000                     0.000


        V13 = 3.554*Self, Errorvar.= 65.967, R² = 0.161
Standerr   (0.576)                   (5.269)
Z-values    6.170                    12.520
P-values    0.000                     0.000


      V17_21 = -0.669*Self, Errorvar.= 38.206, R² = 0.0116
Standerr      (0.371)                   (2.541)
Z-values      -1.806                    15.036
P-values       0.071                     0.000
```

c. Specifically, was the interaction variable, V17_21 statistically significant?

No. So we would re-run the **SIMPLIS** program, dropping the interaction term.

Our results indicate a perfect fit due to a saturated model.

```
Goodness of Fit Statistics
Degrees of Freedom for (C1)-(C2)                                  0
Maximum Likelihood Ratio Chi-Square (C1)     0.0 (P = 1.0000)
Browne's (1984) ADF Chi-Square (C2_NT)       0.00 (P = 1.0000)
        The Model is Saturated, the Fit is Perfect !
Measurement Equations
        V11 = 2.824*Self, Errorvar.= 7.640, R² = 0.511
Standerr   (0.368)                  (1.939)
Z-values    7.677                    3.940
P-values    0.000                    0.000


        V12 = 0.598*Self, Errorvar.= 1.244, R² = 0.224
Standerr   (0.0897)                 (0.118)
Z-values    6.675                   10.581
P-values    0.000                    0.000


        V13 = 3.630*Self, Errorvar.= 65.415, R² = 0.168
Standerr   (0.582)                  (5.313)
Z-values    6.239                   12.312
P-values    0.000                    0.000
```

Graph

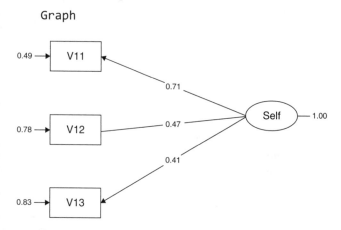

2. LATENT VARIABLE INTERACTION MODEL

An organizational psychologist was investigating whether *work tension* and *collegiality* were predictors of *job satisfaction*. However, research indicated that *work tension* and *collegiality* interact, so an SEM interaction model was hypothesized and tested.

The raw data (jobs.csv) were imported into LISREL, and saved as a LISREL system file, *jobs.lsf*. It includes the 9 observed variables (v1–v9); *job*, *work*, and *colleg* latent variables, and the latent interaction variable, *interact*.

	v1	v2	v3	v4	v5	v6	v7	v8	v9	job	work	colleg	interact
1	3.00	2.00	3.00	4.00	4.00	4.00	2.00	0.00	4.00	0.92	1.47	0.69	1.00
2	2.00	1.00	2.00	3.00	2.00	3.00	0.00	0.00	1.00	-0.08	0.16	-0.73	-0.11
3	2.00	1.00	4.00	2.00	2.00	2.00	0.00	4.00	4.00	0.40	0.14	0.69	0.10
4	1.00	1.00	2.00	2.00	4.00	4.00	3.00	4.00	4.00	-0.08	1.13	1.47	1.66
5	2.00	0.00	1.00	2.00	3.00	2.00	1.00	4.00	4.00	-0.63	0.27	0.92	0.24
6	4.00	3.00	3.00	2.00	4.00	2.00	1.00	3.00	4.00	1.25	0.84	0.85	0.72
7	0.00	0.00	1.00	2.00	1.00	2.00	0.00	2.00	1.00	-1.17	-0.62	-0.55	0.34
8	4.00	2.00	2.00	2.00	2.00	2.00	0.00	4.00	4.00	0.55	0.16	0.70	0.11
9	3.00	3.00	2.00	2.00	3.00	4.00	2.00	4.00	4.00	0.84	0.98	1.24	1.21
10	0.00	3.00	3.00	3.00	1.00	3.00	1.00	4.00	4.00	0.39	0.26	0.90	0.24
11	0.00	0.00	0.00	0.00	0.00	0.00	0.00	0.00	0.00	-1.68	-1.84	-1.40	2.57
12	1.00	0.00	1.00	1.00	1.00	0.00	0.00	1.00	1.00	-1.12	-1.22	-0.85	1.04
13	1.00	1.00	3.00	2.00	2.00	2.00	1.00	2.00	1.00	-0.07	-0.12	-0.28	0.03
14	1.00	1.00	1.00	1.00	1.00	1.00	1.00	1.00	1.00	-0.76	-0.92	-0.61	0.56
15	2.00	2.00	2.00	1.00	2.00	1.00	1.00	1.00	1.00	0.06	-0.49	-0.53	0.26
16	0.00	0.00	0.00	0.00	0.00	0.00	0.00	1.00	1.00	-1.67	-1.76	-0.96	1.68
17	0.00	0.00	0.00	0.00	0.00	0.00	0.00	0.00	0.00	-1.68	-1.84	-1.40	2.57
	0.00	0.00	0.00	0.00	0.00	0.00	1.00	0.00	-1.68	-1.80	-1.23	2.22

Run the SIMPLIS program to analyze the Interaction Model.

```
Latent Interaction Variable Model - No Intercept Term
Observed Variables: v1-v9 job work colleg interact
```

```
Raw Data from File jobs.lsf
Sample Size = 200
Relationships:
job = work colleg interact
Path Diagram
End of Problem
```

The structural equation indicated that no interaction effect is present between *work tension* and *collegiality* ($z = .96$, $p = .337$). Rather, *work tension* and *collegiality* were statistically significant predictors of job satisfaction as direct linear effects. The $R^2 = .801$.

Structural Equations

```
job = 0.984*work - 0.180*colleg + 0.0361*interact, Errorvar.=
   0.219, R² = 0.801
  Standerr  (0.0647)  (0.0783)   (0.0376)   (0.0220)
  Z-values   15.201    -2.299      0.960      9.924
  P-values    0.000     0.021      0.337      0.000
```

The latent interaction variable should be dropped and the SIMPLIS program run again. The $R^2 = .80$, so little change, which indicates that the interaction effect did not contribute to the prediction of job satisfaction.

Structural Equations

```
job = 0.974*work - 0.168*colleg, Errorvar.= 0.220, R² = 0.800
  Standerr  (0.0640)   (0.0773)     (0.0221)
  Z-values   15.234    -2.176        9.950
  P-values    0.000     0.030        0.000
```

ANSWERS TO ODD EXERCISES

CHAPTER 1

1. Define the following terms:

    a. Latent variable: an unobserved variable that is not directly measured, but is computed using multiple observed variables.
    b. Observed variable: a raw score obtained from a test or measurement instrument on a trait of interest.
    c. Dependent variable: a variable that is measured and related to outcomes, performance, or criteria.
    d. Independent variable: a variable that defines mutually exclusive categories, for example, gender, region, or grade level), or is measured as a continuous variable, for example, test scores, and influences a dependent variable.

3. Explain the difference between an independent latent variable and an independent observed variable. An *independent latent variable* is not directly measured, but is computed using multiple independent observed variables. An *independent observed variable* is a raw score obtained from a measurement instrument and used as a predictor variable.

CHAPTER 2

1. Define the following levels of measurement:

    a. Nominal: mutually exclusive groups or categories with number or percentage indicated.
    b. Ordinal: mutually exclusive groups or categories that are ordered with a ranking indicated.

    c.   Interval: continuous data with arbitrary zero point, appropriately permitting a mean and a standard deviation.

    d.   Ratio: continuous data with a true zero point, appropriately permitting a mean and a standard deviation.

3.   Explain how each of the following affects statistics:

    a.   Restriction of range: A set of scores that are restricted in range implies reduced variability. Variance and covariance are important in statistics, especially affecting correlation.

    b.   Missing data: A set of scores with missing data can affect the estimate of the mean and standard deviation. It is important to determine whether the missing data are due to a data entry error, are missing at random, or are missing systematically due to some other variable (for example, gender).

    c.   Outliers: A set of scores with an outlier (extreme score) can affect the estimate of the mean and standard deviation. It is important to determine whether the outlier is an incorrect data value due to data entry error, represents another group of subjects not well sampled, or potentially requires the researcher to gather more data to fill in between the range of existing data.

    d.   Non-linearity: Researchers have generally analyzed relationships in data assuming linearity. Linearity is a requirement for the Pearson correlation coefficient. Consequently, a lack of linearity that is not included in the statistical model would yield misleading results.

    e.   Non-normality: Skewness, or lack of symmetry in the frequency distribution, and kurtosis, the departure from the peakedness of a normal distribution, affect inferential statistics, especially the mean, the standard deviation, and correlation coefficient estimates. Data transformations, especially a probit transformation, can help to yield a more normally distributed set of scores.

CHAPTER 3

1.   Partial and part correlations:

$$r_{12.3} = \frac{.6 - (.7)\,(.4)}{\sqrt{[1 - (.7)^2][1 - (.4)^2]}} = .49$$

$$r_{1(2.3)} = \frac{.6 - (.7)\,(.4)}{\sqrt{[1 - (.4)^2]}} = .35.$$

3. A meaningful theoretical relationship should be plausible given that:

    a. Variables logically precede each other in time.
    b. Variables covary or correlate together as expected.
    c. Other influences or "causes" are controlled.
    d. Variables should be measured on at least an interval level.
    e. Changes in a preceding variable should affect variables that follow, either directly or indirectly.

5. Enter the correlation matrix in a statistics package ($N = 209$). Calculate the determinant of the matrix, eigenvalues, and eigenvectors. For example, R uses det() and eigen() functions (see Rdet.r file). Report and interpret these values.

Correlation Matrix							
Academic	1.00						
Athletic	.43	1.00					
Attract	.50	.48	1.00				
GPA	.49	.22	.32	1.00			
Height	.10	-.04	-.03	.18	1.00		
Weight	.04	.02	-.16	-.10	.34	1.00	
Rating	.09	.14	.43	.15	-.16	-.27	1.00
S. D.	.16	.07	.49	3.49	2.91	19.32	1.01
Means	.12	.05	.42	10.34	.00	94.13	2.65

CHAPTER 4

1. The following LISREL-SIMPLIS program is run to analyze the theoretical regression model for predicting gross national product (GNP) from knowledge of labor, capital, and time:

```
Regression of GNP
Observed variables: GNP LABOR CAPITAL TIME
Covariance Matrix:
 4256.530
  449.016    52.984
 1535.097   139.449   1114.447
  537.482    53.291    170.024 73.747
Sample Size: 23
Equation: GNP = LABOR CAPITAL TIME
Number of Decimals = 3
Path Diagram
End of Problem
```

CHAPTER 5

< Single program Answer>

CHAPTER 6

< Single program Answer>

CHAPTER 7

1. Define the following SEM modeling steps:

   Model specification: Developing a theoretical model to test based on all of the relevant theory, research, and information available.

   Model identification: Determining whether a unique set of parameter estimates can be computed given the sample data contained in the sample covariance matrix S and the theoretical model that produced the implied population covariance matrix Σ.

   Model estimation: Obtaining estimates for each of the parameters specified in the model that produced the implied population covariance matrix Σ. The intent is to obtain parameter estimates that yield a matrix Σ as close as possible to S, our sample covariance matrix of the observed or indicator variables. When elements in the matrix S minus the elements in the matrix Σ equal zero ($S - \Sigma = 0$), then $\chi^2 = 0$, indicating a perfect model fit to the data and all values in S are equal to values in Σ.

   Model testing: Determining how well the sample data fit the theoretical model. In other words, to what extent is the theoretical model supported by the obtained sample data? Global omnibus tests of the fit of the model are available as well as the fit of individual parameters in the model.

   Model modification: Changing the initial implied model and retesting the global fit and individual parameters in the new respecified model. To determine how to modify the model, there are a number of procedures available to guide the adding or dropping of paths in the model so that alternative models can be tested.

3. Define model fit, model comparison, and model parsimony.

   Model fit determines the degree to which the sample variance–covariance data fit the structural equation model.

   Model comparison involves comparing an implied model with a null model (independence model). The null model could also be any model that establishes a baseline for expecting other alternative models to be different.

Model parsimony seeks the minimum number of estimated coefficients required to achieve a specific level of model fit. Basically, an overidentified model is compared with a restricted model.

5. How are modification indices in LISREL-SIMPLIS used?

Modification indices in LISREL-SIMPLIS indicate the amount of change in chi-square that would result if a path was added or dropped.

7. How should a researcher test for the difference between two alternative models?

A researcher computes a chi-square difference test (referred to as a likelihood ratio test) by subtracting the base model chi-square value from the constrained model chi-square value with one degree of freedom. When testing measurement invariance between groups in a specified model, the comparative fit index (CFI) and McDonald's non-centrality index (NCI) are recommended.

9. Explain the four-step approach to modeling in SEM.

The four-step approach first uses exploratory factor analysis to establish a meaningful theoretical model. Next, one conducts a confirmatory factor analysis with a new sample of data. Then, one conducts a test of the structural equation model. Finally, a researcher tests planned hypotheses about free parameters in the model.

11. Use G*Power 3 to calculate power for modified model with NCP = 6.3496 at $p = .05$, $p = .01$, $p = .001$ levels of significance. What happens to power when alpha increases?

Power decreases as alpha increases (power = .73, alpha = .05; power = .50, alpha = .01, and power = .24, alpha = .001).

CHAPTER 8

< Single Program Answer >

CHAPTER 9

< Single program Answer >

CHAPTER 10

< Single program Answer >

CHAPTER 11

< Single program Answer >

CHAPTER 12

< Single program Answer >

CHAPTER 13

< Single program Answer >

CHAPTER 14

< Single program Answer >

CHAPTER 15

< Single program Answer>

CHAPTER 16

< NONE>

INTRODUCTION TO MATRIX OPERATIONS

Structural equation modeling performs calculations using several different matrices. The matrix operations to perform the calculations involve addition, subtraction, multiplication, and division of elements in the different matrices. We present these basic matrix operations, followed by a simple multiple regression example.

Matrix Definition

A matrix is indicated by a capital letter, for example, A, B, or R, and takes the form:

$$A = \begin{bmatrix} 3 & 5 \\ 5 & 6 \end{bmatrix}$$

The matrix can be rectangular or square-shaped and contains an array of numbers. A correlation matrix is a square matrix with the value of 1.0 in the diagonal and variable correlations in the off-diagonal. A correlation matrix is symmetrical because the correlation coefficients in the lower half of the matrix are the same as the correlation coefficients in the upper half of the matrix. We usually only report the diagonal values and the correlations in the lower half of the correlation matrix. For example:

$$R = \begin{bmatrix} 1.0 & .30 & .50 \\ .30 & 1.0 & .60 \\ .50 & .60 & 1.0 \end{bmatrix}$$

but we report the following as a correlation matrix:

```
1.0
 .30 1.0
 .50 .60 1.0
```

Matrices have a certain number of rows and columns. The A matrix above has 2 rows and 2 columns. The **order** of a matrix is the **size** of the matrix, or number of rows times the number of columns. The order of the A matrix is $2 \times 2 = 4$.

We use row and column designations to identify the location of the elements in the matrix. Each element has a two-digit subscript, where the first digit is the row number and the second digit is the column number. For example, the correlation $r = .30$ is in the R_{21} matrix location or row 2, column 1.

Matrix Addition and Subtraction

Matrix addition adds corresponding elements in two matrices, while matrix subtraction subtracts corresponding elements in two matrices. Consequently, the two matrices must have the same order (numbers of rows and columns. In the following example, matrix A elements are added to matrix B elements:

$$A + B = \begin{bmatrix} 3 & 5 & 2 \\ 1 & 6 & 0 \\ 9 & 1 & 2 \end{bmatrix} + \begin{bmatrix} 1 & -3 & 5 \\ 2 & 1 & 3 \\ 0 & 7 & -3 \end{bmatrix} = \begin{bmatrix} 4 & 2 & 7 \\ 3 & 7 & 3 \\ 9 & 8 & -1 \end{bmatrix}$$

Matrix Multiplication

Matrix multiplication is not as straightforward as matrix addition and subtraction. For a product of matrices we write $A \cdot B$ or AB. The number of columns in the first matrix must match the number of rows in the second matrix to be compatible and permit multiplication of the elements of the matrices. If A is an $m \times n$ matrix and B is an $n \times p$ matrix, then AB is an $m \times p$ matrix. The following example will illustrate how the row elements in the first matrix (A) are multiplied by the column elements in the second matrix (B) to yield the elements in the third matrix C.

```
c₁₁ = 1 × 2 + 2 × 1 = 2 + 2 = 4
c₁₂ = 1 × 4 + 2 × 8 = 4 + 16 = 20
c₁₃ = 1 × 6 + 2 × 7 = 6 + 14 = 20
c₂₁ = 3 × 2 + 5 × 1 = 6 + 5 = 11
c₂₂ = 3 × 4 + 5 × 8 = 12 + 40 = 52
c₂₃ = 3 × 6 + 5 × 7 = 18 + 35 = 53
```

$$A \bullet B = \begin{bmatrix} 1 & 2 \\ 3 & 5 \end{bmatrix} \bullet \begin{bmatrix} 2 & 4 & 6 \\ 1 & 8 & 7 \end{bmatrix} = \begin{bmatrix} 4 & 20 & 20 \\ 11 & 52 & 53 \end{bmatrix}$$

Matrix C is:

$$C = \begin{bmatrix} 4 & 20 & 20 \\ 11 & 52 & 53 \end{bmatrix}$$

It is important to note that matrix multiplication is **noncommutative**, i.e., $AB \neq BA$. The order of operation in multiplying elements of the matrices is therefore very important. Matrix multiplication, however, is associative, i.e., $A(BC) = (AB)C$, because the order of matrix multiplication is maintained.

A special matrix multiplication is possible when a single number is multiplied by the elements in a matrix. The single number is called a **scalar**. The scalar is simply multiplied by each of the elements in the matrix. For example,

$$D = 2 \begin{bmatrix} 3 & 4 \\ 4 & 6 \end{bmatrix} = \begin{bmatrix} 6 & 8 \\ 8 & 12 \end{bmatrix}$$

Matrix Division

Matrix division is similar to matrix multiplication with a little twist. In regular division, we divide the numerator by the denominator. However, we can also multiply the numerator by the inverse of the denominator. For example, in regular division, 4 is divided by 2 to give 2; however, we get the same result if we multiply 4 by ½. Therefore, matrix division is simply A / B or $A \bullet 1/B = AB^{-1}$. The B^{-1} matrix is called the **inverse** of the B matrix.

Matrix division thus requires finding the inverse of a matrix, which involves computing the **determinant** of a matrix, the **matrix of minors**, and the **matrix of cofactors**. We then create a **transposed matrix** and an **inverse matrix**, which when multiplied yield an **identity matrix**. We now turn our attention to finding these values and matrices involved in matrix division.

Determinant of a Matrix

The determinant of a matrix is a unique number (not a matrix) that uses all the elements in the matrix for its calculation, and is a generalized variance for that matrix. For our illustration we will compute the determinant of a 2 by 2 matrix, leaving higher-order matrix determinant computations for high-speed computers. The determinant is computed by cross-multiplying the elements of the matrix: if

$$A = \begin{bmatrix} a & b \\ c & d \end{bmatrix}$$

then the determinant of $A = ad - cb$.

For example, when

$$A = \begin{bmatrix} 2 & 5 \\ 3 & 6 \end{bmatrix}$$

the determinant of $A = 2 \times 6 - 3 \times 5 = -3$.

Matrix of Minors

Each element in a matrix has a **minor**. To find the minor of each element, simply draw a vertical and a horizontal line through that element to form a matrix with one less row and column. We next calculate the determinants of these minor matrices, and then place them in a **matrix of minors**. The matrix of minors has the same number of rows and columns as the original matrix.

The **matrix of minors** for the following 3 by 3 matrix is computed as follows:

$$A = \begin{bmatrix} 1 & 6 & -3 \\ -2 & 7 & 1 \\ 3 & -1 & 4 \end{bmatrix}$$

$$M_{11} = \begin{bmatrix} 7 & 1 \\ -1 & 4 \end{bmatrix} = (7)(4) - (1)(-1) = 29$$

$$M_{12} = \begin{bmatrix} -2 & 1 \\ 3 & 4 \end{bmatrix} = (-2)(4) - (1)(3) = -11$$

$$M_{13} = \begin{bmatrix} -2 & 7 \\ 3 & -1 \end{bmatrix} = (-2)(-1) - (7)(3) = -19$$

$$M_{21} = \begin{bmatrix} 6 & -3 \\ -1 & 4 \end{bmatrix} = (6)(4) - (-1)(-3) = 21$$

$$M_{22} = \begin{bmatrix} 1 & -3 \\ 3 & 4 \end{bmatrix} = (1)(4) - (-3)(3) = 13$$

$$M_{23} = \begin{bmatrix} 1 & 6 \\ 3 & -1 \end{bmatrix} = (1)(-1) - (6)(3) = -19$$

$$M_{31} = \begin{bmatrix} 6 & -3 \\ 7 & 1 \end{bmatrix} = (6)(1) - (-3)(7) = 27$$

$$M_{32} = \begin{bmatrix} 1 & -3 \\ -2 & 1 \end{bmatrix} = (1)(1) - (-3)(-2) = -5$$

$$M_{33} = \begin{bmatrix} 1 & 6 \\ -2 & 7 \end{bmatrix} = (1)(7) - (6)(-2) = 19$$

$$A_{Minors} = \begin{bmatrix} 29 & -11 & -19 \\ 21 & 13 & -19 \\ 27 & -5 & 19 \end{bmatrix}$$

Matrix of Cofactors

A **matrix of cofactors** is created by multiplying each element of the **matrix of minors** by $(-1)^{i+j}$, where i = row number of the element and j = column number of the element. These values form a new matrix called a **matrix of cofactors**.

An easy way to remember this multiplication rule is to observe the matrix below. Start with the first row and multiply the first entry by (+), second entry by (–), third by (+), and so on to the end of the row. For the second row start multiplying by (–), then (+), then (–), and so on. All odd rows begin with a – sign and all even rows begin with a + sign.

```
+    -    +
-    +    -
+    -    +
-    +    -
```

So for the matrix above we obtain the **matrix of cofactors**:

$$C_{Cofactors} = \begin{bmatrix} 29 & 11 & -19 \\ -21 & 13 & 19 \\ 27 & 5 & 19 \end{bmatrix}$$

Determinant of Matrix Revisited

The matrix of cofactors makes finding the determinant of any size matrix easy. We multiply elements in any row or column of our original A matrix, by any one corresponding row or column in the **matrix of cofactors** to compute the determinant of the matrix. We can compute the determinant using any row or column, so rows with zeroes make the calculation of the determinant easier. The determinant of our original 3 by 3 matrix (A) using the 3 by 3 matrix of cofactors would be: $\det A = a_{11}c_{11} + a_{12}c_{12} + a_{13}c_{13}$

Recall that matrix A was:

$$A = \begin{bmatrix} 1 & 6 & -3 \\ -2 & 7 & 1 \\ 3 & -1 & 4 \end{bmatrix}$$

The matrix of cofactors was:

$$C_{Cofactors} = \begin{bmatrix} 29 & 11 & -19 \\ -21 & 13 & 19 \\ 27 & 5 & 19 \end{bmatrix}$$

So, the determinant of matrix A, using the first row of both matrices is.

$$\det A = (1)(29) + (6)(11) + (-3)(-19) = 152$$

We could have also used the second columns of both matrices and obtained the same determinant value:

$$\det A = (6)(11) + (7)(13) + (-1)(5) = 152$$

Two special matrices that we have already mentioned also have determinants: a **diagonal matrix** and a **triangular matrix**. A **diagonal matrix** is a matrix which contains zero or non-zero elements on its main diagonal, but zeroes everywhere else. A **triangular matrix** has zeros only either above or below the main diagonal. To calculate the determinants of these matrices, we only need to multiply the elements on the main diagonal. For example, the following triangular matrix K has a determinant of 96.

$$K = \begin{bmatrix} 2 & 0 & 0 & 0 \\ 4 & 1 & 0 & 0 \\ -1 & 5 & 6 & 0 \\ 3 & 9 & -2 & 8 \end{bmatrix}$$

This is computed by multiplying the diagonal values in the matrix:

$$\det K = (2)(1)(6)(8) = 96$$

Transpose of a Matrix

The transpose of a matrix is created by taking the *rows* of an original matrix C and placing them into corresponding *columns* of the transpose matrix, C'. For example:

$$C = \begin{bmatrix} 29 & 11 & -19 \\ -21 & 13 & 19 \\ 27 & 5 & 19 \end{bmatrix}$$

$$C' = \begin{bmatrix} 29 & -21 & 27 \\ 11 & 13 & 5 \\ -19 & 19 & 19 \end{bmatrix}$$

The **transposed matrix** of the **matrix of cofactors** is called the **adjoint matrix**, designated as Adj(A). The **adjoint matrix** is important because we use it to create the inverse of a matrix, our final step in matrix division operations.

Inverse of a Matrix

The general formula for finding an inverse of a matrix is one over the determinant of the matrix times the adjoint of the matrix:

$$A^{-1} = [1/ \det A] \, \text{Adj}(A)$$

Since we have already found the determinant and adjoint of A, we find the inverse of A as follows:

$$A^{-1} = \left(\frac{1}{152}\right) \begin{bmatrix} 29 & -21 & 27 \\ 11 & 13 & 5 \\ -19 & 19 & 19 \end{bmatrix} = \begin{bmatrix} .191 & -.138 & .178 \\ .072 & .086 & .033 \\ -.125 & .125 & .125 \end{bmatrix}$$

An important property of the inverse of a matrix is that if we multiply its elements by the elements in our original matrix, we should obtain an **identity matrix**. An identity matrix will have 1.0 in the diagonal and zeroes in the off-diagonal. The identity matrix is computed as:

$$A \, A^{-1} = I$$

Because we have the original matrix A and the inverse of matrix A, we multiply elements of the matrices to obtain the **identity** matrix, I:

$$A A^{-1} = \begin{bmatrix} 1 & 6 & -3 \\ -2 & 7 & 1 \\ 3 & -1 & 4 \end{bmatrix} \begin{bmatrix} .191 & -.138 & .178 \\ .072 & .086 & .033 \\ -.125 & .125 & .125 \end{bmatrix} = \begin{bmatrix} 1 & 0 & 0 \\ 0 & 1 & 0 \\ 0 & 0 & 1 \end{bmatrix}$$

Matrix Operations in Statistics

We now turn our attention to how the matrix operations are used to compute statistics. We will only cover the calculation of the Pearson correlation and provide the matrix approach in multiple regression, leaving more complicated analyses to computer software programs.

Pearson Correlation (Variance–Covariance Matrix)

In the book, we illustrated how to compute the Pearson correlation coefficient from a variance–covariance matrix. Here we demonstrate the matrix approach. An important matrix in computing correlations is the sums of squares and cross-products matrix (SSCP). We will use the following pairs of scores to create the SSCP matrix.

X1	X2
5	1
4	3
6	5

The mean of X1 is 5 and the mean of X2 is 3. We use these mean values to compute deviation scores from each mean. We first create a matrix of deviation scores, D:

$$D = \begin{bmatrix} 5 & 1 \\ 4 & 3 \\ 6 & 5 \end{bmatrix} - \begin{bmatrix} 5 & 3 \\ 5 & 3 \\ 5 & 3 \end{bmatrix} = \begin{bmatrix} 0 & -2 \\ -1 & 0 \\ 1 & 2 \end{bmatrix}$$

Next, we create the transpose of matrix D, D':

$$D' = \begin{bmatrix} 0 & -1 & 1 \\ -2 & 0 & 2 \end{bmatrix}$$

Finally, we multiply the transpose of matrix D by the matrix of deviation scores to compute the sums of squares and cross-products matrix:

$$SSCP = D' * D$$

$$SSCP = \begin{bmatrix} 0 & -1 & 1 \\ -2 & 0 & 2 \end{bmatrix} * \begin{bmatrix} 0 & -2 \\ -1 & 0 \\ 1 & 2 \end{bmatrix} = \begin{bmatrix} 2 & 2 \\ 2 & 8 \end{bmatrix}$$

The sums of squares are along the diagonal of the matrix, and the sum of squares cross-products are on the off-diagonal. The matrix multiplications are provided below for the interested reader.

$(0)(0) + (-1)(-1) + (1)(1) = 2$ [sums of squares $= (0^2 + -1^2 + 1^2)$]

$(-2)(0) + (0)(-1) + (2)(1) = 2$ [sum of squares cross-product]

$(0)(-2) + (-1)(0) + (1)(2) = 2$ [sum of squares cross-product]

$(-2)(-2) + (0)(0) + (2)(2) = 8$ [sums of squares $= (-2^2 + 0^2 + 2^2)$]

$SSCP = \begin{bmatrix} 2 & 2 \\ 2 & 8 \end{bmatrix}$ Sum of squares in diagonal of matrix

Variance–Covariance Matrix

Structural equation modeling uses a sample variance–covariance matrix in its calculations. The SSCP matrix is used to create the variance–covariance matrix, S:

$$S = \frac{SSCP}{n-1}$$

In matrix notation this becomes ½ times the matrix elements:

Covariance terms in the off-diagonal of matrix

$S = \frac{1}{2} * \begin{bmatrix} 2 & 2 \\ 2 & 8 \end{bmatrix} = \begin{bmatrix} 1 & 1 \\ 1 & 4 \end{bmatrix}$ Variance of variables in diagonal of matrix

We can now calculate the Pearson correlation coefficient using the basic formula of covariance divided by the square root of the product of the variances.

$$r = \frac{Covariance X1 X2}{\sqrt{Variance X1 * Variance X2}} = \frac{1}{\sqrt{1 * 4}} = \frac{1}{2} = .5$$

Multiple Regression

The multiple linear regression equation with two predictor variables is:

$$y = \beta_0 + \beta_1 x_1 + \beta_2 x_2 + e_i$$

where y is the dependent variable, x_1 and x_2 the two predictor variables,

β_0 is the regression constant or y-intercept,
β_1 and β_2 are the regression weights to be estimated,
and e is the error of prediction.

Given the data below, we can use matrix algebra to estimate the regression weights:

y	x_1	x_2
3	2	1
2	3	5
4	5	3
5	7	6
8	8	7

We model each subject's y score as a linear function of the betas:

$$y_1 = 3 = 1\beta_0 + 2\beta_1 + 1\beta_2 + e_1$$
$$y_2 = 2 = 1\beta_0 + 3\beta_1 + 5\beta_2 + e_2$$
$$y_3 = 4 = 1\beta_0 + 5\beta_1 + 3\beta_2 + e_3$$
$$y_4 = 5 = 1\beta_0 + 7\beta_1 + 6\beta_2 + e_4$$
$$y_5 = 8 = 1\beta_0 + 8\beta_1 + 7\beta_2 + e_5$$

This series of equations can be expressed as a single matrix equation:

$$y = \begin{bmatrix} 3 \\ 2 \\ 4 \\ 5 \\ 8 \end{bmatrix} = \begin{bmatrix} 1 & 2 & 1 \\ 1 & 3 & 5 \\ 1 & 5 & 3 \\ 1 & 7 & 6 \\ 1 & 8 & 7 \end{bmatrix} \begin{bmatrix} \beta_0 \\ \beta_1 \\ \beta_2 \end{bmatrix} + \begin{bmatrix} e_1 \\ e_2 \\ e_3 \\ e_4 \\ e_5 \end{bmatrix}$$

The first column of matrix X contains 1s, which compute the regression constant. In matrix form, the multiple linear regression equation is: $y = X\beta + e$.

Using calculus, we translate this matrix to solve for the regression weights:

$$\hat{\beta} = (X'X)^{-1}X'y$$

The matrix equation is:

$$\hat{\beta} = \left\{ \begin{bmatrix} 1 & 1 & 1 & 1 & 1 \\ 2 & 3 & 5 & 7 & 8 \\ 1 & 5 & 3 & 6 & 7 \end{bmatrix} \overset{X}{\begin{bmatrix} 1 & 2 & 1 \\ 1 & 3 & 5 \\ 1 & 5 & 3 \\ 1 & 7 & 6 \\ 1 & 8 & 7 \end{bmatrix}}^{-1} \right\} * \overset{X'}{\begin{bmatrix} 1 & 1 & 1 & 1 & 1 \\ 2 & 3 & 5 & 7 & 8 \\ 1 & 5 & 3 & 6 & 7 \end{bmatrix}} \overset{y}{\begin{bmatrix} 3 \\ 2 \\ 4 \\ 5 \\ 8 \end{bmatrix}}$$

We first compute $X'X$ and then compute $X'y$

$$X'X = \begin{bmatrix} 5 & 25 & 22 \\ 25 & 151 & 130 \\ 22 & 130 & 120 \end{bmatrix} \quad \text{and} \quad X'y = \begin{bmatrix} 22 \\ 131 \\ 111 \end{bmatrix}$$

Next we create the inverse of $X'X$, where 1016 is the determinant of X'.

$$(X'X)^{-1} = \frac{1}{1016} \begin{bmatrix} 1220 & -140 & -72 \\ -140 & 116 & -100 \\ -72 & -100 & 130 \end{bmatrix}$$

Finally, we solve for the X_1 and X_2 regression weights:

$$\hat{\beta} = \frac{1}{1016} \begin{bmatrix} 1220 & -140 & -72 \\ -140 & 116 & -100 \\ -72 & -100 & 130 \end{bmatrix} \begin{bmatrix} 22 \\ 131 \\ 111 \end{bmatrix} = \begin{bmatrix} .50 \\ 1 \\ -.25 \end{bmatrix}$$

The multiple regression equation is:

$$\hat{y}_i = .50 + 1X_1 - .25 X_2$$

We use the multiple regression equation to compute predicted scores and then compare the predicted values to the original y values to compute the error of prediction values, e. For example, the first y score was 3 with $X_1 = 2$ and $X_2 = 1$. We substitute the $X1$ and $X2$ values in the regression equation and compute a predicted y score of 2.25. The error of prediction is computed as $y -$ this predicted y score or $3 - 2.25 = .75$. These computations are listed below and are repeated for the remaining y values.

$$\hat{y}_1 = .50 + 1(2) - .25(1)$$
$$\hat{y}_1 = 2.25$$
$$\hat{e}_1 = 3 - 2.25 = .75$$

$$\hat{y}_2 = .50 + 1(3) - .25(5)$$
$$\hat{y}_2 = 2.25$$
$$\hat{e}_2 = 2 - 2.25 = -.25$$

$$\hat{y}_3 = .50 + 1(5) - .25(3)$$
$$\hat{y}_3 = 4.75$$
$$\hat{e}_3 = 4 - 4.75 = -.75$$

$$\hat{y}_4 = .50 + 1(7) - .25(6)$$
$$\hat{y}_4 = 6.00$$
$$\hat{e}_4 = 5 - 6 = -1.00$$

$$\hat{y}_5 = .50 + 1(8) - .25(7)$$
$$\hat{y}_5 = 6.75$$
$$\hat{e}_5 = 8 - 6.75 = 1.25$$

The regression equation is: $\hat{y}_i = .50 + 1.0X_1 - .25 X_2$

We can now place the Y values, X values, regression weights, and error terms back into the matrices to yield a complete solution for the Y values. Notice that the error term vector should sum to zero (0.0). Also notice that each y value is uniquely composed of an intercept term (.50), a regression weight (1.0) times an X_1 value, a regression weight (–.25) times an X_2 value, and a residual error, for example, the first y value of $3 = .5 + 1.0(2) - .25(1) + .75$.

$$\begin{bmatrix} 3 \\ 2 \\ 4 \\ 5 \\ 8 \end{bmatrix} = .5 + 1.0 \begin{bmatrix} 2 \\ 3 \\ 5 \\ 7 \\ 8 \end{bmatrix} - .25 \begin{bmatrix} 1 \\ 5 \\ 3 \\ 6 \\ 7 \end{bmatrix} + \begin{bmatrix} .75 \\ -.25 \\ -.75 \\ -1.00 \\ 1.25 \end{bmatrix}$$

STATISTICAL TABLES

Table A.1: Areas Under the Normal Curve (z-scores)

Second decimal place in z

z	.00	.01	.02	.03	.04	.05	.06	.07	.08	.09
.0	.0000	.0040	.0080	.0120	.0160	.0199	.0239	.0279	.0319	.0359
.1	.0398	.0438	.0478	.0517	.0557	.0596	.0636	.0675	.0714	.0753
.2	.0793	.0832	.0871	.0910	.0948	.0987	.1026	.1064	.1103	.1141
.3	.1179	.1217	.1255	.1293	.1331	.1368	.1406	.1443	.1480	.1517
.4	.1554	.1591	.1628	.1664	.1700	.1736	.1772	.1808	.1844	.1879
.5	.1915	.1950	.1985	.2019	.2054	.2088	.2123	.2157	.2190	.2224
.6	.2257	.2291	.2324	.2357	.2389	.2422	.2454	.2486	.2517	.2549
.7	.2580	.2611	.2642	.2673	.2704	.2734	.2764	.2794	.2823	.2852
.8	.2881	.2910	.2939	.2967	.2995	.3023	.3051	.3078	.3106	.3133
.9	.3159	.3186	.3212	.3238	.3264	.3289	.3315	.3340	.3365	.3389
1.0	.3413	.3438	.3461	.3485	.3508	.3531	.3554	.3577	.3599	.3621
1.1	.3643	.3665	.3686	.3708	.3729	.3749	.3770	.3790	.3810	.3830
1.2	.3849	.3869	.3888	.3907	.3925	.3944	.3962	.3980	.3997	.4015
1.3	.4032	.4049	.4066	.4082	.4099	.4115	.4131	.4147	.4162	.4177
1.4	.4192	.4207	.4222	.4236	.4251	.4265	.4279	.4292	.4306	.4319
1.5	.4332	.4345	.4357	.4793	.4382	.4394	.4406	.4418	.4429	.4441
1.6	.4452	.4463	.4474	.4484	.4495	.4505	.4515	.4525	.4535	.4545
1.7	.4554	.4564	.4573	.4582	.4591	.4599	.4608	.4616	.4625	.4633
1.8	.4641	.4649	.4656	.4664	.4671	.4678	.4686	.4693	.4699	.4706
1.9	.4713	.4719	.4726	.4732	.4738	.4744	.4750	.4756	.4761	.4767
2.0	.4772	.4778	.4783	.4788	.4793	.4798	.4803	.4808	.4812	.4817
2.1	.4821	.4826	.4830	.4834	.4838	.4842	.4846	.4850	.4854	.4857
2.2	.4861	.4826	.4868	.4871	.4875	.4878	.4881	.4884	.4887	.4890
2.3	.4893	.4896	.4898	.4901	.4904	.4906	.4909	.4911	.4913	.4916
2.4	.4918	.4920	.4922	.4925	.4927	.4929	.4931	.4932	.4934	.4936
2.5	.4938	.4940	.4941	.4943	.4945	.4946	.4948	.4949	.4951	.4952
2.6	.4953	.4955	.4956	.4957	.4959	.4960	.4961	.4962	.4963	.4964
2.7	.4965	.4966	.4967	.4968	.4969	.4970	.4971	.4972	.4973	.4974
2.8	.4974	.4975	.4976	.4977	.4977	.4978	.4979	.4979	.4980	.4981
2.9	.4981	.4982	.4982	.4983	.4984	.4984	.4985	.4985	.4986	.4986
3.0	.4987	.4987	.4987	.4988	.4988	.4989	.4989	.4989	.4990	.4990
3.1	.4990	.4991	.4991	.4991	.4992	.4922	.4992	.4992	.4993	.4993
3.2	.4993	.4993	.4994	.4994	.4994	.4994	.4994	.4995	.4995	.4995
3.3	.4995	.4995	.4995	.4996	.4996	.4996	.4996	.4996	.4996	.4997
3.4	.4997	.4997	.4997	.4997	.4997	.4997	.4997	.4997	.4997	.4998
3.5	.4998									
4.0	.49997									
4.5	.499997									
5.0	.4999997									

Table A.2: Distribution of *t* for Given Probability Levels

df	Level of significance for one-tailed test					
	.10	*.05*	.025	.01	.005	.0005
	Level of significance for two-tailed test					
	.20	.10	.05	.02	.01	.001
1	3.078	6.314	12.706	31.821	63.657	636.619
2	1.886	2.920	4.303	6.965	9.925	31.598
3	1.638	2.353	3.182	4.541	5.841	12.941
4	1.533	2.132	2.776	3.747	4.604	8.610
5	1.476	2.015	2.571	3.365	4.032	6.859
6	1.440	1.943	2.447	3.143	3.707	5.959
7	1.415	1.895	2.365	2.998	3.499	5.405
8	1.397	1.860	2.306	2.896	3.355	5.041
9	1.383	1.833	2.262	2.821	3.250	4.781
10	1.372	1.812	2.228	2.764	3.169	4.587
11	1.363	1.796	2.201	2.718	3.106	4.437
12	1.356	1.782	2.179	2.681	3.055	4.318
13	1.350	1.771	2.160	2.650	3.012	4.221
14	1.345	1.761	2.145	2.624	2.977	4.140
15	1.341	1.753	2.131	2.602	2.947	4.073
16	1.337	1.746	2.120	2.583	2.921	4.015
17	1.333	1.740	2.110	2.567	2.898	3.965
18	1.330	1.734	2.101	2.552	2.878	3.992
19	1.328	1.729	2.093	2.539	2.861	3.883
20	1.325	1.725	2.086	2.528	2.845	3.850
21	1.323	1.721	2.080	2.518	2.831	3.819
22	1.321	1.717	2.074	2.508	2.819	3.792
23	1.319	1.714	2.069	2.500	2.807	3.767
24	1.318	1.711	2.064	2.492	2.797	3.745
25	1.316	1.708	2.060	2.485	2.787	3.725
26	1.315	1.706	2.056	2.479	2.779	3.707
27	1.314	1.703	2.052	2.473	2.771	3.690
28	1.313	1.701	2.048	2.467	2.763	3.674
29	1.311	1.699	2.045	2.462	2.756	3.659
30	1.310	1.697	2.042	2.457	2.750	3.646
40	1.303	1.684	2.021	2.423	2.704	3.551
60	1.296	1.671	2.000	2.390	2.660	3.460
120	1.289	1.658	1.980	2.358	2.617	3.373
∞	1.282	1.645	1.960	2.326	2.576	3.291

Table A.3: Distribution of *r* for Given Probability Levels

df	Level of significance for one-tailed test			
	.05	.025	.01	.005
	Level of significance for two-tailed test			
	.10	.05	.02	.01
1	.988	.997	.9995	.9999
2	.900	.950	.980	.990
3	.805	.878	.934	.959
4	.729	.811	.882	.917
5	.669	.754	.833	.874
6	.622	.707	.789	.834
7	.582	.666	.750	.798
8	.540	.632	.716	.765
9	.521	.602	.685	.735
10	.497	.576	.658	.708
11	.576	.553	.634	.684
12	.458	.532	.612	.661
13	.441	.514	.592	.641
14	.426	.497	.574	.623
15	.412	.482	.558	.606
16	.400	.468	.542	.590
17	.389	.456	.528	.575
18	.378	.444	.516	.561
19	.369	.433	.503	.549
20	.360	.423	.492	.537
21	.352	.413	.482	.526
22	.344	.404	.472	.515
23	.337	.396	.462	.505
24	.330	.388	.453	.496
25	.323	.381	.445	.487
26	.317	.374	.437	.479
27	.311	.367	.430	.471
28	.306	.361	.423	.463
29	.301	.355	.416	.486
30	.296	.349	.409	.449
35	.275	.325	.381	.418
40	.257	.304	.358	.393
45	.243	.288	.338	.372
50	.231	.273	.322	.354
60	.211	.250	.295	.325
70	.195	.232	.274	.303
80	.183	.217	.256	.283
90	.173	.205	.242	.267
100	.164	.195	.230	.254

Table A.4: Distribution of Chi-square for Given Probability Levels

df	Probability .99	.98	.95	.90	.80	.70	.50	.30	.20	.10	.05	.02	.01	.001
1	.00016	.00063	.00393	.0158	.0642	.148	.455	1.074	1.642	2.706	3.841	5.412	6.635	10.827
2	.0201	.0404	.103	.211	.446	.713	1.386	2.408	3.219	4.605	5.991	7.824	9.210	13.815
3	.115	.185	.352	.584	1.005	1.424	2.366	3.665	4.642	6.251	7.815	9.837	11.345	16.266
4	.297	.429	.711	1.064	1.649	2.195	3.357	4.878	5.989	7.779	9.488	11.668	13.277	18.467
5	.554	.752	1.145	1.610	2.343	3.000	4.351	6.064	7.289	9.236	11.070	13.388	15.086	20.515
6	.872	1.134	1.635	2.204	3.070	3.828	5.348	7.231	8.558	10.645	12.592	15.033	16.812	22.457
7	1.239	1.564	2.167	2.833	3.822	4.671	6.346	8.383	9.803	12.017	14.067	16.622	18.475	24.322
8	1.646	2.032	2.733	3.490	4.594	5.527	7.344	9.524	11.030	13.362	15.507	18.168	20.090	26.125
9	2.088	2.532	3.325	4.168	5.380	6.393	8.343	10.656	12.242	14.684	16.919	19.679	21.666	27.877
10	2.558	3.059	3.940	4.865	6.179	7.267	9.342	11.781	13.442	15.987	18.307	21.161	23.209	29.588
11	3.053	3.609	4.575	5.578	6.989	8.148	10.341	12.899	14.631	17.275	19.675	22.618	24.725	31.264
12	3.571	4.178	5.226	6.304	7.807	9.034	11.340	14.011	15.812	18.549	21.026	24.054	26.217	32.909
13	4.107	4.765	5.892	7.042	8.634	9.926	12.340	15.119	16.985	19.812	22.362	25.472	27.688	34.528
14	4.660	5.368	6.571	7.790	9.467	10.821	13.339	16.222	18.151	21.064	23.685	26.873	29.141	36.123
15	5.229	5.985	7.261	8.547	10.307	11.721	14.339	17.322	19.311	22.307	24.996	28.259	30.578	37.697
16	5.812	6.614	7.962	9.312	11.152	12.624	15.338	18.418	20.465	23.542	26.296	29.633	32.000	39.252
17	6.408	7.255	8.672	10.085	12.002	13.531	16.338	19.511	21.615	24.769	27.587	30.995	33.409	40.790
18	7.015	7.906	9.390	10.865	12.857	14.440	17.338	20.601	22.760	25.989	28.869	32.346	34.805	42.312
19	7.633	8.567	10.117	11.651	13.716	15.352	18.338	21.689	23.900	27.204	30.144	33.687	36.191	43.820
20	8.260	9.237	10.851	12.443	14.578	16.266	19.337	22.775	25.038	28.412	31.410	35.020	37.566	45.315
21	8.897	9.915	11.591	13.240	15.445	17.182	20.337	23.858	26.171	29.615	32.671	36.343	38.932	46.797
22	9.542	10.600	12.338	14.041	16.314	18.101	21.337	24.939	27.301	30.813	33.924	37.659	40.289	48.268

Note: For larger values of df, the expression $\sqrt{(X^2)^2} - \sqrt{2df - 1}$ may be used as a normal deviate with unit variance, remembering that the probability for X^2 corresponds with that of a single tail of the normal curve.

Table A.4 Distribution of Chi-square for Given Probability Levels (Continued)

Probability

df	.99	.98	.95	.90	.80	.70	.50	.30	.20	.10	.05	.02	.01	.001
23	10.196	11.293	13.091	14.848	17.187	19.021	22.337	26.018	28.429	32.007	35.172	38.968	41.638	49.728
24	10.856	11.992	13.848	15.659	18.062	19.943	23.337	27.096	29.553	33.196	36.415	40.270	42.980	51.179
25	11.524	12.697	14.611	16.473	18.940	20.867	24.337	28.172	30.675	34.382	37.652	41.566	44.314	52.620
26	12.198	13.409	15.379	17.292	19.820	21.792	25.336	29.246	31.795	35.563	38.885	42.856	45.642	54.052
27	12.879	14.125	16.151	18.114	20.703	22.719	26.336	30.319	32.912	36.741	40.113	44.140	46.963	55.476
28	13.565	14.847	16.928	18.939	21.588	23.647	27.336	31.391	34.027	37.916	41.337	45.419	48.278	56.893
29	14.256	15.574	17.708	19.768	22.475	24.577	28.336	32.461	35.139	39.087	42.557	46.693	49.588	58.302
30	14.953	16.306	18.493	20.599	23.364	25.508	29.336	33.530	36.250	40.256	43.773	47.962	50.892	59.703
32	16.362	17.783	20.072	22.271	25.148	27.373	31.336	35.665	38.466	42.585	46.194	50.487	53.486	62.487
34	17.789	19.275	21.664	23.952	26.938	29.242	33.336	37.795	40.676	44.903	48.602	52.995	56.061	65.247
36	19.233	20.783	23.269	25.643	28.735	31.115	35.336	39.922	42.879	47.212	50.999	55.489	58.619	67.985
38	20.691	22.304	24.884	27.343	30.537	32.992	37.335	42.045	45.076	49.513	53.384	57.969	61.162	70.703
40	22.164	23.838	26.509	29.051	32.345	34.872	39.335	44.165	47.269	51.805	55.759	60.436	63.691	73.402
42	23.650	25.383	28.144	30.765	34.147	36.755	41.335	46.282	49.456	54.090	58.124	62.892	66.206	76.084
44	25.148	26.939	29.787	32.487	35.974	38.641	43.335	48.396	51.639	56.369	60.481	65.337	68.710	78.750
46	26.657	28.504	31.439	34.215	37.795	40.529	45.335	50.507	53.818	58.641	62.830	67.771	71.201	81.400
48	28.177	30.080	33.098	35.949	39.621	42.420	47.335	52.616	55.993	60.907	65.171	70.197	73.683	84.037
50	29.707	31.664	34.764	37.689	41.449	44.313	49.335	54.723	58.164	63.167	67.505	72.613	76.154	86.661
52	31.246	33.256	36.437	39.433	43.281	46.209	51.335	56.827	60.332	65.422	69.832	75.021	78.616	89.272
54	32.793	34.856	38.116	41.183	45.117	48.106	53.335	58.930	62.496	67.673	72.153	77.422	81.069	91.872
56	34.350	36.464	39.801	42.937	46.955	50.005	55.335	61.031	64.658	69.919	74.468	79.815	83.513	94.461
58	35.913	38.078	41.492	44.696	48.797	51.906	57.335	63.129	66.816	72.160	76.778	82.201	85.950	97.039
60	37.485	39.699	43.188	46.459	50.641	53.809	59.335	65.227	68.972	74.397	79.082	84.580	88.379	99.607
62	39.063	41.327	44.889	48.226	52.487	55.714	61.335	67.322	71.125	76.630	81.381	86.953	90.802	102.166
64	40.649	42.960	46.595	49.996	54.336	57.620	63.335	69.416	73.276	78.860	83.675	89.320	93.217	104.716
66	42.240	44.599	48.305	51.770	56.188	59.527	65.335	71.508	75.424	81.085	85.965	91.681	95.626	107.258
68	43.838	46.244	50.020	53.548	58.042	61.436	67.335	73.600	77.571	83.308	88.250	94.037	98.028	109.791
70	45.442	47.893	51.739	55.329	59.898	63.346	69.335	75.689	79.715	85.527	90.531	96.388	100.425	112.317

Table A.5 Distribution of *F* for Given Probability Levels (.05 Level)

df$_1$ df$_2$	1	2	3	4	5	6	7	8	9	10	12	15	20	24	30	40	60	120	∞
1	161.40	199.50	215.70	224.60	230.20	234.00	236.80	238.90	240.50	241.90	243.90	245.90	248.00	249.10	250.10	251.10	252.20	253.30	254.30
2	18.51	19.00	19.16	19.25	19.30	19.33	19.35	19.37	19.38	19.49	19.41	19.43	19.45	19.45	19.46	19.47	19.48	19.49	19.50
3	10.13	9.55	9.28	9.12	9.01	8.94	8.89	8.85	8.81	8.79	8.74	8.70	8.66	8.64	8.62	8.59	8.57	8.55	8.53
4	7.71	6.94	6.59	6.39	6.26	6.15	6.09	6.04	6.00	5.96	5.91	5.86	5.80	5.77	5.75	5.72	5.69	5.66	5.63
5	6.61	5.79	5.41	5.19	5.05	4.95	4.88	4.82	4.77	4.74	4.68	4.62	4.56	4.53	4.50	4.46	4.43	4.40	4.36
6	5.99	5.14	4.76	4.53	4.39	4.28	4.21	4.15	4.10	4.06	4.00	3.94	3.87	3.84	3.81	3.77	3.74	3.70	3.67
7	5.59	4.74	4.35	4.12	3.97	3.87	3.79	3.73	3.68	3.64	3.57	3.51	3.44	3.41	3.38	3.34	3.30	3.27	3.23
8	5.32	4.46	4.07	3.84	3.69	3.58	3.50	3.44	3.39	3.35	3.28	3.22	3.15	3.12	3.08	3.04	3.01	2.97	2.93
9	5.12	4.26	3.86	3.63	3.48	3.37	3.29	3.23	3.18	3.14	3.07	3.01	2.94	2.90	2.86	2.83	2.79	2.75	2.71
10	4.96	4.10	3.71	3.48	3.33	3.22	3.14	3.07	3.02	2.98	2.91	2.85	2.77	2.74	2.70	2.66	2.62	2.58	2.54
11	4.84	3.98	3.59	3.36	3.20	3.09	3.01	2.95	2.90	2.85	2.79	2.72	2.65	2.61	2.57	2.53	2.49	2.45	2.40
12	4.75	3.89	3.49	3.26	3.11	3.00	2.91	2.85	2.80	2.75	2.69	2.62	2.54	2.51	2.47	2.43	2.38	2.34	2.30
13	4.67	3.81	3.41	3.18	3.03	2.92	2.83	2.77	2.71	2.67	2.60	2.53	2.46	2.42	2.38	2.34	2.30	2.25	2.21
14	4.60	3.74	3.34	3.11	2.96	2.85	2.76	2.70	2.65	2.60	2.53	2.46	2.39	2.35	2.31	2.27	2.22	2.18	2.13
15	4.54	3.68	3.29	3.06	2.90	2.79	2.71	2.64	2.59	2.54	2.48	2.40	2.33	2.29	2.25	2.20	2.16	2.11	2.07
16	4.49	3.63	3.24	3.01	2.85	2.74	2.66	2.59	2.54	2.49	2.42	2.35	2.28	2.24	2.19	2.15	2.11	2.06	2.01
17	4.45	3.59	3.20	2.96	2.81	2.70	2.61	2.55	2.49	2.45	2.38	2.31	2.23	2.19	2.15	2.10	2.06	2.01	1.96
18	4.41	3.55	3.16	2.93	2.77	2.66	2.58	2.51	2.46	2.41	2.34	2.27	2.19	2.15	2.11	2.06	2.02	1.97	1.92
19	4.38	3.52	3.13	2.90	2.74	2.63	2.54	2.48	2.42	2.38	2.31	2.23	2.16	2.11	2.07	2.03	1.98	1.93	1.88
20	4.35	3.49	3.10	2.87	2.71	2.60	2.51	2.45	2.39	2.35	2.28	2.20	2.12	2.08	2.04	1.99	1.95	1.90	1.84
21	4.32	3.47	3.07	2.84	2.68	2.57	2.49	2.42	2.37	2.32	2.25	2.18	2.10	2.05	2.01	1.96	1.92	1.87	1.81
22	4.30	3.44	3.05	2.82	2.66	2.55	2.46	2.40	2.34	2.30	2.23	2.15	2.07	2.03	1.98	1.94	1.89	1.84	1.78
23	4.28	3.42	3.03	2.80	2.64	2.53	2.44	2.37	2.32	2.27	2.20	2.13	2.05	2.01	1.96	1.91	1.86	1.81	1.76
24	4.26	3.40	3.01	2.78	2.62	2.51	2.42	2.36	2.30	2.25	2.18	2.11	2.03	1.98	1.94	1.89	1.84	1.79	1.73
25	4.24	3.39	2.99	2.76	2.60	2.49	2.40	2.34	2.28	2.24	2.16	2.09	2.01	1.96	1.92	1.87	1.82	1.77	1.71
26	4.23	3.37	2.98	2.74	2.59	2.47	2.39	2.32	2.27	2.22	2.15	2.07	1.99	1.95	1.90	1.85	1.80	1.75	1.69
27	4.21	3.35	2.96	2.73	2.57	2.46	2.37	2.31	2.25	2.20	2.13	2.06	1.97	1.93	1.88	1.84	1.79	1.73	1.67
28	4.20	3.34	2.95	2.71	2.56	2.45	2.36	2.29	2.24	2.19	2.12	2.04	1.96	1.91	1.87	1.82	1.77	1.71	1.65
29	4.18	3.33	2.93	2.70	2.55	2.43	2.35	2.28	2.22	2.18	2.10	2.03	1.94	1.90	1.85	1.81	1.75	1.70	1.64

Table A.5 Distribution of F for Given Probability Levels (.05 Level) (Continued)

df_1/df_2	1	2	3	4	5	6	7	8	9	10	12	15	20	24	30	40	60	120	∞
30	4.17	3.32	2.92	2.69	2.53	2.42	2.33	2.27	2.21	2.16	2.09	2.01	1.93	1.89	1.84	1.79	1.74	1.68	1.62
40	4.08	3.23	2.84	2.61	2.45	2.34	2.25	2.18	2.12	2.08	2.00	1.92	1.84	1.79	1.74	1.69	1.64	1.58	1.51
60	4.00	3.15	2.76	2.53	2.37	2.25	2.17	2.10	2.04	1.99	1.92	1.84	1.75	1.70	1.65	1.59	1.53	1.47	1.39
120	3.92	3.07	2.68	2.45	2.29	2.17	2.09	2.02	1.96	1.91	1.83	1.75	1.66	1.61	1.55	1.50	1.43	1.35	1.25
∞	3.84	3.00	2.60	2.37	2.21	2.10	2.01	1.94	1.88	1.83	1.75	1.67	1.57	1.52	1.46	1.39	1.32	1.22	1.00

Table A.6 Distribution of F for Given Probability Levels (.01 Level)

df_2 \ df_1	1	2	3	4	5	6	7	8	9	10	12	15	20	24	30	40	60	120	∞
1	4052.00	4999.50	5403.00	5625.00	5764.00	5859.00	5928.00	5982.00	6022.00	6056.00	6106.00	6157.00	6209.00	6235.00	6261.00	6287.00	6313.00	6339.00	6366.00
2	98.5	99.00	99.17	99.25	99.30	99.33	99.36	99.37	99.39	99.40	99.42	99.43	99.45	99.46	99.47	99.47	99.48	99.49	99.50
3	34.12	30.82	29.46	28.71	28.24	27.91	27.67	27.49	27.25	27.23	27.05	26.87	26.69	26.60	26.50	26.41	26.32	26.22	26.13
4	21.20	18.00	16.69	15.98	15.52	15.21	14.98	14.80	14.66	14.55	14.37	14.20	14.02	13.93	13.84	13.75	13.65	13.56	13.46
5	16.26	13.27	12.06	11.39	10.97	10.67	10.46	10.29	10.16	10.05	9.89	9.72	9.55	9.47	9.38	9.29	9.20	9.11	9.02
6	13.75	10.92	9.78	9.15	8.75	8.47	8.26	8.10	7.98	7.87	7.72	7.56	7.40	7.31	7.23	7.14	7.06	6.97	6.88
7	12.25	9.55	8.45	7.85	7.46	7.19	6.99	6.84	6.72	6.62	6.47	6.31	6.16	6.07	5.99	5.91	5.82	5.74	5.65
8	11.26	8.65	7.59	7.01	6.63	6.37	6.18	6.03	5.91	5.81	5.67	5.52	5.36	5.28	5.20	5.12	5.03	4.95	4.86
9	10.56	8.02	6.99	6.42	6.06	5.80	5.61	5.47	5.35	5.26	5.11	4.96	4.81	4.73	4.65	4.57	4.48	4.40	4.31
10	10.04	7.56	6.55	5.99	5.64	5.39	5.20	5.06	4.94	4.85	4.71	4.56	4.41	4.33	4.25	4.17	4.08	4.00	3.91
11	9.65	7.21	6.22	5.67	5.32	5.07	4.89	4.74	4.63	4.54	4.40	4.25	4.10	4.02	3.94	3.86	3.78	3.69	3.60
12	9.33	6.93	5.95	5.41	5.06	4.82	4.64	4.50	4.39	4.30	4.16	4.01	3.86	3.78	3.70	3.62	3.54	3.45	3.36
13	9.07	6.70	5.74	5.21	4.86	4.62	4.44	4.30	4.19	4.10	3.96	3.82	3.66	3.59	3.51	3.43	3.34	3.25	3.17
14	8.86	6.51	5.56	5.04	4.69	4.46	4.28	4.14	4.03	3.94	3.80	3.66	3.51	3.43	3.35	3.27	3.18	3.09	3.00
15	8.68	6.36	5.42	4.89	4.56	4.32	4.14	4.00	3.89	3.80	3.67	3.52	3.37	3.29	3.21	3.13	3.05	2.96	2.87
16	8.53	6.23	5.29	4.77	4.44	4.20	4.03	3.89	3.78	3.69	3.55	3.41	3.26	3.18	3.10	3.02	2.93	2.84	2.75
17	8.40	6.11	5.18	4.67	4.34	4.10	3.93	3.79	3.68	3.59	3.46	3.31	3.16	3.08	3.00	2.92	2.83	2.75	2.65
18	8.29	6.01	5.09	4.58	4.25	4.01	3.84	3.71	3.60	3.51	3.37	3.23	3.08	3.00	2.92	2.84	2.75	2.66	2.57
19	8.18	5.93	5.01	4.50	4.17	3.94	3.77	3.63	3.52	3.43	3.30	3.15	3.00	2.92	2.84	2.76	2.67	2.58	2.49
20	8.10	5.85	4.94	4.43	4.10	3.87	3.70	3.56	3.46	3.37	3.23	3.09	2.94	2.86	2.78	2.69	2.61	2.52	2.42
21	8.02	5.78	4.87	4.37	4.04	3.81	3.64	3.51	3.40	3.31	3.17	3.03	2.88	2.80	2.72	2.64	2.55	2.46	2.36
22	7.95	5.72	4.82	4.31	3.9	3.76	3.59	3.45	3.35	3.26	3.12	2.98	2.83	2.75	2.67	2.58	2.50	2.40	2.31
23	7.88	5.66	4.76	4.26	3.94	3.71	3.54	3.41	3.30	3.21	3.07	2.93	2.78	2.70	2.62	2.54	2.45	2.35	2.26
24	7.82	5.61	4.72	4.22	3.90	3.67	3.50	3.36	3.26	3.17	3.03	2.89	2.74	2.66	2.58	2.49	2.40	2.31	2.21
25	7.77	5.57	4.68	4.18	3.85	3.63	3.46	3.32	3.22	3.13	2.99	2.85	2.70	2.62	2.54	2.45	2.36	2.27	2.17
26	7.72	5.53	4.64	4.14	3.82	3.59	3.42	3.29	3.18	3.09	2.96	2.81	2.66	2.58	2.50	2.42	2.33	2.23	2.13
27	7.68	5.49	4.60	4.11	3.78	3.56	3.39	3.26	3.15	3.06	2.93	2.78	2.63	2.55	2.47	2.38	2.29	2.20	2.10

Table A.6 Distribution of F for Given Probability Levels (.01 Level) (Continued)

df_2 \ df_1	1	2	3	4	5	6	7	8	9	10	12	15	20	24	30	40	60	120	∞
28	7.64	5.45	4.57	4.07	3.75	3.53	3.36	3.23	3.12	3.03	2.90	2.75	2.60	2.52	2.44	2.35	2.26	2.17	2.06
29	7.60	5.42	4.54	4.04	3.73	3.50	3.33	3.20	3.09	3.00	2.87	2.73	2.57	2.49	2.41	2.33	2.23	2.14	2.03
30	7.56	5.39	4.51	4.02	3.70	3.47	3.30	3.17	3.07	2.98	2.84	2.70	2.55	2.47	2.39	2.30	2.21	2.11	2.01
40	7.31	5.18	4.31	3.83	3.51	3.29	3.12	2.99	2.89	2.80	2.66	2.52	2.37	2.29	2.20	2.11	2.02	1.92	1.80
60	7.08	4.98	4.13	3.65	3.34	3.12	2.95	2.82	2.72	2.63	2.50	2.35	2.20	2.12	2.03	1.94	1.84	1.73	1.60
120	6.85	4.79	3.95	3.48	3.17	2.96	2.79	2.66	2.56	2.47	2.34	2.19	2.03	1.95	1.86	1.76	1.66	1.53	1.38
∞	6.63	4.61	3.78	3.32	3.02	2.80	2.64	2.51	2.41	2.32	2.18	2.04	1.88	1.79	1.70	1.59	1.47	1.32	1.00

NAME INDEX

SUBJECT INDEX

Page entries with a 't' denote tables.